IMMUNOLOGY AND IMMUNE SYSTEM DISORDERS

IMMUNE MODULATION BY PROBIOTICS

STATE OF THE ART

IMMUNOLOGY AND IMMUNE SYSTEM DISORDERS

Additional books in this series can be found on Nova's website
under the Series tab.

Additional e-books in this series can be found on Nova's website
under the e-book tab.

IMMUNOLOGY AND IMMUNE SYSTEM DISORDERS

IMMUNE MODULATION BY PROBIOTICS

STATE OF THE ART

ALESSANDRO DI CERBO
BENIAMINO PALMIERI
AND
MARIA APONTE

NOVA BIOMEDICAL

New York

Library of Congress Cataloging-in-Publication Data

ISBN: 978-1-62948-954-4

Library of Congress Control Number: 2014932382

Published by Nova Science Publishers, Inc. † New York

Contents

Abstract

Recent evidences suggest that both gut and colonic microbiota are critical for the development of the immune system and are the major modulators of the mucosal immune system.

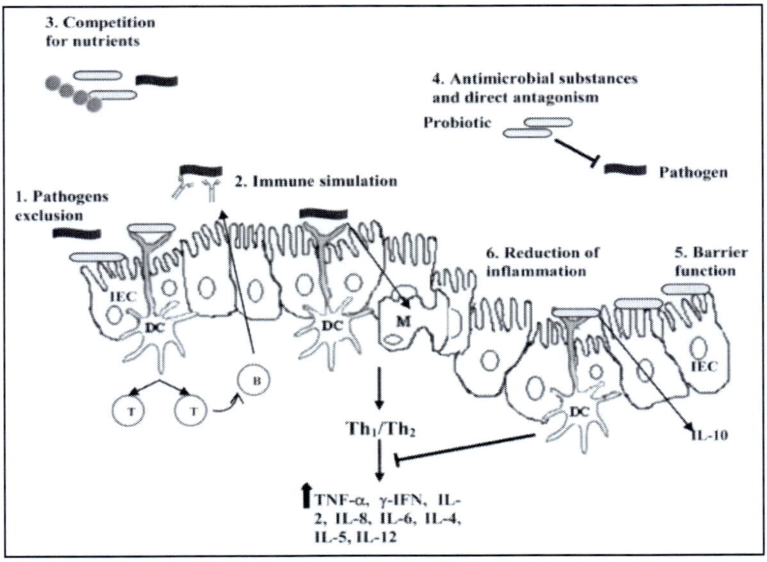

Figure 1. Some probiotic mechanisms whereby induce several beneficial host responses. Most effect consist on: 1) Exclusion and competing with pathogen to epithelial cells adhesion, 2) innate immune stimulation, 3) competition for nutrients and prebiotic products, 4) production of antimicrobial substances and thereby pathogen antagonism, 5) protection of intestinal barrier integrity and 6) regulation of anti-inflammatory cytokine and inhibition of pro-inflammatory cytokine production. IEC: intestinal epithelium cells, DC: dendritic cell; IL: interleukin; M: intestinal M cell [1].

Current literature suggests that manipulation of such microbiota through supplementation with probiotics may significantly affect the host's immune responses (Figure 1).

Bifidobacteria and Lactobacilli, that are components of the commensal flora, have been associated with gut homeostasis, competitive exclusion of pathogens, production of antimicrobial compounds, promotion of gut barrier function, and immune modulation. It has been postulated that early inclusion of lactobacilli in the intestinal microbiota (Figure 1), during neonatal colonization or by probiotic supplementation, might prevent the development of allergic disorders. Nowadays the intestinal microflora is considered a postnatally acquired organ exposed to environmental influences and modulated by external factors such as nutrition.

Acknowledgments

This work was supported by a grant from the Province of Avellino.

Chapter I

Introduction

The recognition of bacteria as non-self agents by mammalian cells has a key role in producing innate response to control infection. Several bacterial antigens are perceived as “non self” molecules by host immune cells, using receptors of the innate immune system. PAMPs (pathogen-associated molecular patterns) are found in both Gram-negative and Gram-positive bacteria, (lipopeptides, peptidoglycan, flagellin, and bacterial DNA).

Others are specific for Gram-positive bacteria (lipoteichoic acid) [2]. Toll-like receptors (TLRs) belong to a family of leucine-rich repeat (LRR) proteins that are either expressed at the cell surface or in intracellular compartments [3]. They recognize and bind a wide variety of bacterial PAMPs, such as LPS (TLR4, even if some LPS species can be recognized by TLR2), lipopeptides (ricognized by TLR1, TLR2 and TLR6), lipoarabinomannan and lipoteichoic acid (TLR2), flagellin (TLR5), and bacterial DNA (TLR9)[4-6].

The binding of PAMPs to TLRs induces a deep and complex reset of immune cells and activates several innate pathways such as inflammation, coagulation, and cell death. Some of the nucleotide-binding oligomerization domain (NOD) LRR proteins, such as Nod1 and Nod2, are also intracellular sensors for bacterial molecules and mediate inflammation on recognition of specific motifs in bacterial peptidoglycan [7]. It has been observed that CD14, a GPI-anchored glycoprotein, binds PAMPs, associates with TLRs, and markedly enhances cell responsiveness to several bacterial molecules such as LPS and LAM [8].

Chapter II

Chronology of Probiotics Modulated Immunity Studies

Historical Background

Yasui et al., (1989) compared the gut intraluminal immunogenicity of *Bifidobacterium breve*, included in fermented milk with that of Bacteroides thetaiotaomicron [9]. *In vivo*, serum antibody to *Bifidobacterium breve* was detected first in mice fed the organism for 33 days; antibody decreased in mice fed these for more than 33 days. Serum antibody to *Bacteroides thetaiotaomicron* was detected in mice fed for 7 days and was long-term maintained at the same level in mice fed these for more than 7 days. From *in vitro* tests, the optimal doses of *Bifidobacterium breve* and *Bacteroides thetaiotaomicron* to induce antibody production by Peyer's patch cells, intestinal lymphoid tissue cells, were 5 x 10^8 and 5 x 10^7 bacteria/mL, respectively. Therefore, it was suggested that immunogenicity of *Bifidobacterium breve* was weaker than that of *Bacteroides thetaiotaomicron*. Furthermore, the change of specific antibody production by Peyer's patch cells in the mice administered *Bifidobacterium breve* orally was tested by the Peyer's patch cell culture method. Antibody production against *Bifidobacterium breve* by Peyer's patch cells in mice given *Bifidobacterium breve* for 25 and for 33 days increased and decreased, respectively, in comparison with the control. These results suggested that when serum antibody to *Bifidobacterium breve* increased significantly, anti-*Bifidobacterium breve* antibody production by Peyer's patch cells was suppressed, and thereafter, serum antibodies to *Bifidobacterium breve* decreased and were no more detected. These findings supported the hypothesis that serum antibody production to *Bifidobacterium breve* was regulated in Peyer's patches. Two years later the influence of formaldehyde-killed *Escherichia coli* strain Nissle 1917 (SK 22) on macrophages of C57BL/6 mice *in vitro* was investigated by Hockerz [10]. SK 22 activated macrophages derived from bone marrow produced Interleukin-6 with high efficiency. In addition, SK 22 stimulated macrophages to secrete tumor necrosis factor. Furthermore, macrophages were activated by SK 22 to produce a 3 fold amount of oxygen radicals compared to the spontaneous oxygen radical production. In contrast to this finding, the phagocytic capacity of these macrophages was only slightly increased. The specific lysis of P 815 tumor cells by peritoneal macrophages after coincubation with SK 22 strengthened the concept that SK 22 was an efficient immunomodulator of the unspecific immune system. The immuneregulating activity of probiotics was evaluted also for those used in food processing [11]. Solis Pereyra et al., (1991) evaluated the production of interferon by fasted human subjects in response to *Lactobacillus bulgaricus* and *Streptococcus thermophilus in vivo* and *in vitro.* The 2'-5' A synthetase activity of blood mononuclear cells was used to estimate interferon production following a single ingestion of 10^{11} bacteria in yoghurt or sterile milk (controls). The level of the 2'-5' A synthetase of the yoghurt fed subjects was 83% (p = 0.002) higher than that of the milk fed controls 24 hours after ingestion. The baseline value remained unchanged in the control group. Blood mononuclear cells from a second group of subjects, were cultured with lactic bacteria for 48 hours, their cell-free supernatants contained gamma interferon, thus suggesting a transient production of interferon.

Kaila et al., (1992) randomly assigned either a *Lactobacillus rhamnosus* GG fermented milk product (study group) or a pasteurized yogurt (placebo group) to 39 children with acute rotavirus [12]. The diarrhea follow up was significantly shorter in the study group than in the placebo group [mean 1.1 (SD 0.6) versus 2.5 (SD 1.4) d, p = 0.001]. *Lactobacillus* GG therapy was associated with a significantly enhanced nonspecific humoral response during the acute phase of the infection, reflected in the IgG, IgA, and IgM Ig-secreting cell numbers. In the remission phase, 90% of the study group versus 46% of the placebo group developed an IgA specific antibody-secreting cell response to rotavirus (p = 0.006). The results indicated that *Lactobacillus* GG promoted recovery from rotavirus diarrhea enhancing the local immune defense. Furthermore, specific IgA response to rotavirus was endorsed, suggesting a protective role of *Lactobacillus* GG against reinfections.

Johansson et al., (1993), whereas, studied both colonization by different lactobacilli (19 strains in a fermented oatmeal soup at a concentration of 5 x 10^6 CFU/g for 10 days) on human intestinal mucosa of healthy volunteers and their on different groups of indigenous bacteria [13]. They took biopsies from both the upper jejunum (figure 2 A) and rectum (figure 2 B) 1 day before administration and 1 and 11 days after administration ending. The administration significantly increased the lactobacilli counts on jejunum mucosa and their levels remained high even 11 days after administration ending.

In 1994, Perdigón et al. studied the effect yogurt supplementation to BALB/c mice on the various gut-associated lymphoid cells [14]. Throughout the feeding period, 2, 5, 7 and 10 consecutive days, there was a marked cell infiltration of plasma cells and lymphocytes.

The number of macrophages on the small intestine increased significantly after feeding for 2 d, while the β-glucuronidase activity was only slightly higher compared with the controls. After a 7 d feeding period, the number of IgA secreting cells increased, while the values for cells secreting IgM and IgG and for T lymphocytes remained similar to those of the controls. The yogurt administration effect on lymphoid cells associated with the large intestine achieved a definite increase of the IgA secreting B cells and T lymphocytes, along the whole feeding period especially in the latter type of cell. Since giving yogurt mainly enhanced the IgA secreting B cells in both small and large intestines, this increase would strengthen the host's defence mechanisms in the intestinal mucosa. Both immune response and intestinal flora modulation were also investigated by fermented milk, containing *Lactobacillus acidophilus* La1 and bifidobacteria, administration [15]. Volunteers drank fermented milk over a period of three weeks during which an attenuated *Salmonella typhi* Ty21a was administered to mimic an enteropathogenic infection. A control group ate no fermented foods but received the *Salmonella typhi* Ty21a. Faecal flora analyses showed an increase in *Lactobacillus acidophilus* and bifidobacterial counts during fermented milk intake. The specific serum IgA titre to *Salmonella typhi* Ty21a in the test group was > 4-fold and significantly higher (p = 0.04) than in the control group. An increase in total serum IgA was also observed. Lactic acid bacteria (LAB) which can persist in the gastrointestinal tract acted as adjuvants to the humoral immune response.

(A)	MEDIAN BACTERIAL COUNTS (log CFU/g of mucosa) in upper jejuna			(B)	MEDIAN BACTERIAL COUNTS (log CFU/g of mucosa) in recta		
GROUP	BEFORE ADMINISTRATION (n = 12)	1 DAY AFTER ADMINISTRATION ENDED (n = 12)	11 DAYS AFTER ADMINISTRATION ENDED (n = 10)	GROUP	BEFORE ADMINISTRATION (n = 12)	1 DAY AFTER ADMINISTRATION ENDED (n = 12)	11 DAYS AFTER ADMINISTRATION ENDED (n = 10)
Anaerobic bacteria	4,1	4,6	4,6	Anaerobic bacteria	6,6	6,4	6,0
Aerobic bacteria	3,9	4,3	4,5	Aerobic bacteria	6,4	5,6	5,6
Gram-negative anaerobic bacteria	3,8	3,4	3,1	Gram-negative anaerobic bacteria	6,0	5,7	4,9
Gram-positive anaerobic bacteria	3,8	4,1	4,2	Gram-positive anaerobic bacteria	6,3	6,1	5,8
Gram-positive aerobic bacteria	4,1	3,9	3,8	Gram-positive aerobic bacteria	6,5	5,4	5,8
Lactobacillus spp.	3,0	3,9	4,0	*Lactobacillus spp.*	4,6	5,2	5,3
Lactic acid bacteria	3,8	4,3	4,4	Lactic acid bacteria	4,9	4,7	4,9
Anaerobic bacteria on azide blood agar	3,4	4,2	4,4	Anaerobic bacteria on azide blood agar	5,3	4,5	4,7
Aerobic bacteria on azide blood agar	3,6	4,0	4,0	Aerobic bacteria on azide blood agar	4,0	4,4	3,6
Sulfite-reducing clostridia	3,1	3,5	-	Sulfite-reducing clostridia	4,9	4,2	5,0
Enterococcus spp.	3,8	4,2	4,4	*Enterococcus spp.*	3,6	4,0	3,4
Enterobacteriaceae	-	-	-	*Enterobacteriaceae*	6,1	5,3	5,4

Figure 2. Bacterial counts in upper jejuna at the three sampling times (A); Bacterial counts in recta at the three sampling times (B). (Modified from Johansson et al., 1993) [13].

Aso et al., (1995) evaluated the recurrence prophylaxis by an oral *Lactobacillus casei* preparation (BLP) on 138 patients affected by superficial transitional bladder cell carcinoma following transurethral resection [16]. Patients were stratified into the following 3 subgroups: (A) with primary multiple tumors, (B) with recurrent single tumors, and (C) with recurrent multiple tumors. In each group, patients were randomly allocated to receive BLP or placebo. BLP showed a better prophylactic effect in subgroups A and B than placebo, whereas no significant difference was observed in subgroup C. The outcome with BLP was significantly better than with placebo ($p = 0.01$). Slight and tolerable adverse reactions occurred in three patients receiving BLP and in 3 of the placebo-treated patients. Oral administration of BLP was safe and effective in preventing recurrence of superficial bladder cancer.

In 1995, Schiffrin et al. administered either *Lactobacillus acidophilus* strain La1 or *Bifidobacterium bifidum* strain Bb12 for 3 weeks on healthy volunteers [17]. Blood samples were collected throughout the study to assess changes in lymphocyte subsets or leukocyte phagocytic activity following consumption of the fermented products. No modifications of lymphocyte subpopulations were detected but phagocytosis of *Escherichia coli* sp. *in vitro* was enhanced after the administration of both fermented products. The phagocytosis index enhancement started with fecal colonization by the LAB and was detected up to 6 weeks after ingestion of the fermented products. The fecal lactobacilli and bifidobacteria concentration returned to the baseline prior to consumption. Nonspecific, anti-infective mechanisms of defense could be enhanced by the ingestion of specific LAB strains improving the immune function of particular age groups, such as the neonate or the elderly, for which these functions are diminished. An interesting experiment was undertaken by Mikes and coworkers, which administered a lyophilized *Enterococcus faecium* M-74 to 12 adult subjects in a daily oral dose of 5×10^9 bacteria for six weeks [18]. The bacterium temporarily colonized the host intestine and was recovered in the stools 6 weeks after the last administration. The mean levels of serum cholesterol and LDL showed a biphasic effect - an elevation followed by a sharp decrease (on day 64 of investigation). The decrease was positively matched with a significant increase in the ability to reduce iodonitrotetrazolium and superoxide production by peripheral neutrophils incubated with zymosan or phorbol myristate acetate, and also with an elevated production of IgG by peripheral blood mononuclear cells. Hence, intake of *Enterococcus faecium* might induce a hypocholesterolemic and immunostimulatory effect. Furthermore, *Enterococcus faecium* significantly reduced the average activity of β-D-glucuronidase in stools.

Isolauri et al., (1995) tested the effect of orally administered *Lactobacillus casei* strain GG in conjunction with D x RRV rhesus-human reassortant live oral rotavirus vaccine in 2-5-month-old infants [19]. Infants who received LGG showed an increased response with regard to rotavirus-specific IgM secreting cells on day 8 after vaccination. In infants receiving Lactobacillus casei strain GG or placebo, respectively, a rotavirus IgM seroconversion was detected in 26/27 (96%), versus 23/27 (85%) cases (p = 0.15) and rotavirus IgA seroconversion was detected in 26/28 (93%) versus 20/27 (74%) cases (p = 0.05). These findings suggested that *Lactobacillus casei* strain GG had an immunostimulating effect on oral rotavirus vaccination. The immune responseto rotavirus in children with acute specific gastroenteritis was also investigated by Majamaa et al., (1995) who compared different LAB [20]. After initial oral rehydration, 49 children aged 6 to 35 months with rotavirus gastroenteritis randomly received either *Lactobacillus casei* subsp. *casei* strain GG (LGG), *Lactobacillus casei* subsp. *rhamnosus* (Lactophilus), or a combination of *Streptococcus thermophilus* and *Lactobacillus delbrüeckii* subsp. *bulgaricus* (Yalacta) twice daily for 5 days. Serum antibodies to rotavirus, total number of immunoglobulin-secreting cells (ISC), and specific antibody-secreting cells (sASC) to rotavirus were measured at the acute stage and at convalescence. The mean (SD) duration of diarrhea was 1.8 (0.8) days in children who received LGG, 2.8 (1.2) days in those receiving Lactophilus, and 2.6 (1.4) days in those receiving Yalacta (F = 3.3, p = 0.04). The ISC response was comparable in the three study groups, but the rotavirus-specific immune responses were different. *Lactobacillus casei* subsp. *casei* strain GG therapy was associated with an enhancement of IgA sASC to rotavirus and serum IgA antibody level at convalescent stage.

Lactobacillus casei subsp. *casei* strain GG, promoted serum and intestinal immune responses to rotavirus, and thus could be important in establishing immunity against rotavirus reinfections. Interestingly, Kaila et al., (1995) observed that rotavirus serum IgA enzyme immunoassay antibody responses were higher in infants treated with viable *Lactobacillus casei* strain GG than in those treated with inactivated *Lactobacillus casei* strain GG [21]. There was a significant difference at convalescence with rotavirus specific IgA secreting cells found in 10/12 infants receiving viable but only 2/13 infants receiving inactivated *Lactobacillus casei* strain GG (figure 3).

The results indicated that viable *Lactobacillus casei* strain GG stimulated rotavirus specific IgA antibody responses, theoretically significant in the prevention of reinfections.

	Group A	Group B	p value
Specific antibody secreting cells	(No with < 0.5 x 10^6 cells/n. of patients)		
Acute	2/9	2/8	0.66*
Convalescent	10/12	2/13	0.002*
Mean (95% CI) antibodies (EIU)			
Acute	0.04 (0.01-0.03)	0.1 (0.01-0.5)	0.52**
Convalescent	50.7 (29.4-87.2)	22.4 (11.7-43.0)	0.04**

Figure 3. Number of patients showing a rotavirus specific IgA specific antibody secreting cell response, and the level of rotavirus specific IgA serum antibodies at the acute and convalescent stage of rotavirus gastroenteritis, in infants receiving viable (group A) or inactivated (group B) strain GG. (Modified from Kaila et al., 1995) [21].

Undelying conditions	N. of cases	Source of isolate	Antimicrobial therapy	Clinical outcome	Ref.
Clinical renal failure	1	peritoneal dialysis fluid	Erythromicin and Rifampicin x 1 wk	cure	[23]
Acute leukaemia	3	blood	Imipenem	cure	[24]
Chronic myeloid leukaemia, pneumonia	1	sputum	Erythromicin x 3 wks Lincomycin x 1 wk	cure	[25]
Endocarditis (native valve)	1	blood	Ampicillin Gentamicin, x 6 wks	cure	[26]
Endocarditis	1	blood	Penicillin + Tobranycin Amp + Netilmicin	died after replace-ment of mitral valve	[27]
Endocarditis	1	blood	Penicillin, Streptomycin x 4wks (1st episode) Penicillin x 6wks (2nd episode)	cure	[28]
Arteritis	1	blood, pus from ruptured aortic graft	Ampicillin, Gentamicin x 2 wks + Amoxycillin x 4 wks	cure	[29]
Endocarditis	1	blood, excised aortic valve	not stated		[30]
Arteritis	1	blood	nil	died before start of therapy	[29]

Figure 4. Clinically relevant isolates of *Lactobacillus rhamnosus*. (Modified from Kalima et al., 1996) [22].

In 1996, Kalima et al. reported a case, as well as others (figure 4), where *Lactobacillus rhamnosus* isolated from pericardial effusion and blood in a 5-year-old boy with a 2 year history of severe aplastic anaemia and who received a matched, unrelated donor bone marrow transplant [22].

Miettinen et al., (1996) investigated the role of cytokines in interactions between LAB and bifidobacteria (*Bifidobacterium longum* E505, *Lactobacillus paracasei* subsp. *paracasei* E506, *Lactobacillus acidophilus* E507, *Bifidobacterium animalis* E508, *Lactobacillus rhamnosus* E509, *Lactobacillus paracasei* subsp. *paracasei* E510, *Lactobacillus rhamnosus* GG E522 (ATCC 53103), *Lactococcus lactis* subsp. *cremoris* E523 (ARH 74), *Lactococcus lactis* subsp. *lactis* E414 and *Lactobacillus plantarum* E98) and the immune system, measuring production of TNF-α, IL-6, and IL-10 from human peripheral blood mononuclear cells after stimulation with live or glutaraldehyde-fixed bacteria [31]. Production of TNF-α, IL-6, and, in some cases, IL-10 was induced in amounts even greater than those obtained with lipopolysaccharide as a stimulant, suggesting that LAB could stimulate nonspecific immunity.

In 1996, Cleveland et al. used the human monocytic cell line THP-1 to study IL-12 induction by gram-positive bacteria. Lipoteichoic acid (LTA) a predominant surface glycolipid of gram-positive bacteria (such as lactobacilli) potently induced IL-12 p40 gene expression [32]. The fact that LPS inhibited LTA-induced IL-12 production also suggested a common pathway for LPS and LTA in IL-12 activation. Pretreatment of cells with anti-CD14 monoclonal antibody blocked both LPS and LTA induction of IL-12 p40 expression. LTA also induced Th1 development in naive CD4 T cells by an IL-12-dependent mechanism, indicating direct induction of physiologic levels of IL-12.

Saxelin et al., (1996) studied the prevalence of bacteremia due to *Lactobacillus* species during a 4-year period (1989-1992) in southern Finland, which has a population of about 2.5 million [33]. Among 3,317 blood culture isolates, lactobacilli were identified in eight patients, five of whom had a severe disease predisposing to bacteremic complications. The eight strains isolated were identified to the species level and typed by carbohydrate fermentation tests and by direct sequencing of enzymatically amplified 16S rRNA. The results did not provide evidence that any particular species or subspecies of *Lactobacillus* spp. was the cause of the infections; no infections caused by isolates similar to the recently introduced dairy probiotic strain, *Lactobacillus* GG (ATCC 53103), were observed. Infrequent association of lactobacilli with bacteremic infections was observed in spite of the ubiquitous presence of these organisms in the gastrointestinal tract and their widespread consumption in fermented milks; thus, there was strong evidence that their pathogenic potential was very low. Kishi et al., (1996), comparing two kind of bacterial load (600 and 300 million bacteria/day) on two different groups, observed that oral administration of live *Lactobacillus brevis* subsp. coagulans had a statistically significant increase in IFN-α production at 2 weeks ($p < 0.05$) and at 4 weeks ($p < 0.05$) in the group receiving 600 million bacteria/day and at 4 weeks ($p < 0.05$) in the group receiving 300 million bacteria/day [34]. In particular, IFN-α production, in those with initially low levels, rose significantly when either 300 million or 600 million bacteria/day were ingested. Consumption of heat-killed *Lactobacillus brevis* subsp. *coagulans* 300 million bacteria/day did not result in a statistically significant change in IFN-α production. The level of 2-5A synthetase activity remained the same in the control and experimental groups. Thus, oral administration of live *Lactobacillus brevis* subsp. *coagulans* significantly increased IFN-α production in a dose-dependent manner. *Lactobacillus brevis* subsp. *coagulans* intake tended to be most beneficial in subjects with initially low levels of IFN-α production. Heat-treated *Lactobacillus brevis* subsp. *coagulans* did not elicit a response similar to that of the live bacteria.

Larvol et al., (1996) reported a case of liver abscess due to *Lactobacillus acidophilus* in a 39 year-old man with chronic pancreatitis complicated by both endocrine and exocrine insufficiency, and with a choledoco-duodenostomy [35].

Lactobacillus acidophilus was isolated in blood and liver samples. Complete recovery of the liver abscess occurred after antibiotherapy. It was suggested that the abnormally low duodenal pH secondary to pancreatic insufficiency could have promoted both *Lactobacillus acidophilus* adhesion and multiplication in this patient. The choledoco-duodenostomy may then have promoted biliary tract colonization.

Conversely, Malin et al., (1996) indirectly evaluated the intestinal microflora in juvenile chronic arthritis (JCA) by analyzing enzyme activities (urease, β-glucosidase and β-glucuronidase) in faeces [36]. In 18 out of 26 JCA patients, the illness had been diagnosed during the past year. The control group was composed of eight age-matched control patients and 18 family members of JCA patients (3-36 yr). The mean [95% confidence interval (CI)] urease activity, but not the activities of β-glucosidase and *β-glucuronidase*, in faeces from the JCA group differed from that in the control group: 32.3 (26.6-38.1) nmol/min/mg protein vs 24.0 (16.8-31.6), $p = 0.07$ (figure 5 A-B).

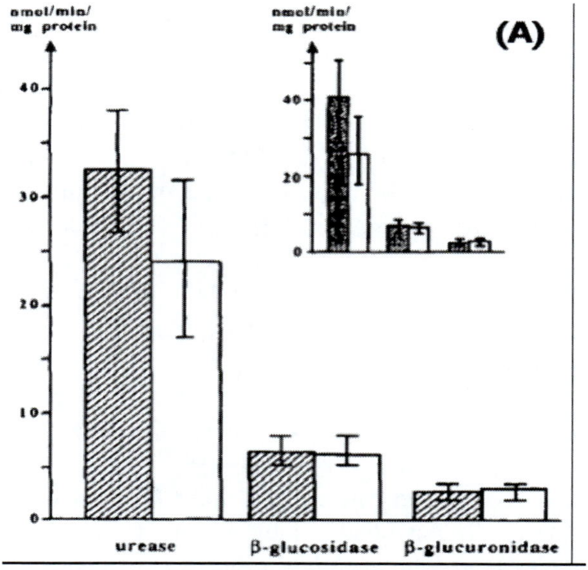

Medication	Urease	β-Glucosidase	β-Glucuronidas
NSAID			
+(n = 22)†	30.0 (23.8–36.0)*	6.1 (4.9–7.3)*	2.4 (1.5–3.3)*
−(n = 4)	45.8 (25.1–66.5)	9.6 (2.8–16.3)	4.6 (2.2–6.9)
Hydroxychloroquine			
+(n = 12)	30.7 (21.2–40.3)	7.5 (5.7–9.2)	3.0 (1.5–4.5)
−(n = 14)	32.4 (23.5–41.2)	5.8 (3.9–7.7)	2.5 (1.5–3.4)
Gold			
+(n = 5)	38.2 (26.8–49.6)	5.7 (3.7–7.6)	18 (0.3–3.3)
−(n = 21)	30.1 (22.9–37.2)	6.8 (5.3–8.3)	2.9 (2.0–3.9)
Sulphasalazine			
+(n = 5)	31.6 (11.6–51.6)	4.6 (0.1–9.2)	3.3 (0.4–6.2)
−(n = 21)	31.6 (24.8–38.5)	7.0 (5.8–8.3)	2.6 (1.7–3.4)

(B)

Figure 5. (a) The mean (95% CI) activity of urease (t = 1.80, $p = 0.07$), β-glucosidase (t = 0.04, $p = 0.97$) and β-glucuronidase (t = 0.05, $p = 0.96$) in faeces in the JCA group, n = 26 (hatched column), and the control group, n = 26 (open column), in the large panel. The small panel shows separately faecal enzyme activities from JCA patients, n = 18 (hatched column), compared to family members, n = 18 (open column): urease (t = 2.30, $p = 0.03$); (b) The mean (95% CI) enzyme activities (nmol/min/mg protein) in faeces from 26 JCA patients with (+) or without (-) NSA1D or anti-rheumatic medication. One patient could have one or more medications [36].

The difference was more marked in a comparison of JCA patients with family members ($p = 0.03$). In a subgroup of subjects, the effect of 10 days oral bacteriotherapy with *Lactobacillus* GG on faecal enzyme activities was then investigated (n = 8 JCA patients, n = 8 control patients). This short-term oral bacteriotherapy reduced the increased urease activity in faeces of JCA patients. Notwithstanding the small number of patients results suggested that the increased urease activity in JCA was specific for the disease, being altered the intestinal microflora in JCA.

Bacteremia by *Lactobacillus* spp. was also evaluated by Antony et al., (1996) in a cohort of 55 patients [37]. Bacteremia with *Lactobacillus* alone occurred in 34 (62%) of the patients, and 12 (22%) of the patients had bacteremia with other organisms, including *Lactobacillus* strains. *Lactobacillus* was isolated from another site in 18 (33%) of these patients. Intravenous catheter infections were not noted in these patients. Underlying conditions included cancer (6 patients), organ transplantation (9), diabetes mellitus (4), and recent surgery (12). Fever occurred in all patients, and eight (15%) of the patients experienced a sepsis syndrome. The mortality rate was 14%; however, only three deaths were attributed only to *Lactobacillus* sepsis. *Lactobacillus* bacteremia is an uncommon condition that usually occurs in patients with severe underlying illnesses and is frequently seen as a part of a polymicrobial infection. Blood cultures positive for lactobacilli represent true infection and not contamination. Although resistance to commonly used antibiotics is common, the mortality rate associated with this bacteremia appears to be low.

Sutas et al., (1996) investigated whether bovine caseins degraded by enzymes of an intestinal bacterial strain, *Lactobacillus* GG (ATCC 53103), could regulate the cytokine production by anti-CD3 antibody-induced peripheral blood mononuclear cells of 14 atopic patients, aged 5-29 (mean, 16) months [38]. Purified casein up-regulated the interleukin-4 and INF-γ production, $p = 0.008$ and $p = 0.008$, respectively. Conversely, *Lactobacillus* GG-degraded casein down-regulated the interleukin-4 production, $p = 0.003$, with no effect on INF-γ. Results indicated that intestinal bacteria could modify immunomodulatory properties of native food proteins and introduced a promising tool to provide protection from potentially harmful dietary antigens at a young age.

Fourty-five patients with clinically significant *Lactobacillus* bacteremia occurring over 15 years were further evaluated by Husni et al., (1997) [39]. Underlying conditions were common, including cancer (40%), recent surgery (38%), and diabetes mellitus (27%). Twenty-two patients were in the intensive care unit at the time of onset of *Lactobacillus* bacteremia (figure 6). 11 of the 45 patients were receiving immunosuppressive therapy, 11 were receiving total parenteral nutrition, and 23 had received antibiotics without activity against *Lactobacillus* prior to the occurrence of bacteremia.

Case no., classification	Patient's age (y)/sex	Diagnosis	Underlying condition(s) (treatment)	Organism(s) isolated	Therapy	Outcome (microbiological/clinical)
Definite						
1	46/F	Bacteremia	Lung transplantation (P/A/C), SLE (?), pulmonary fibrosis	*Lactobacillus* species	Imi, Cm × 4 d	Cured/died
2	70/F	Bacteremia	CABG, CRF, *Clostridium difficile* colitis, DVT, TPN	*Lactobacillus* species	Cm × 8 d	Cured/died
3	56/M	Bacteremia	Sarcoidosis (prednisone)	*Lactobacillus casei*	Gm × 2 d	Cured/died
4	64/F	Intraabdominal abscess*	Diabetes mellitus, colonic cancer, sigmoidectomy, liver cirrhosis, TPN, broad-spectrum ABT	*Lactobacillus* species, *Enterococcus* species	Ctax × 3 d	Cured/died
5	72/M	Bacteremia*	Diabetes mellitus, CVA, liver adenocancer, CAD and valvular disease, GI bleeding	*Lactobacillus* species, *Bacillus* species	Vm × 12 d	Cured/died
6	66/F	Bacteremia	Common bile duct carcinoma, CAD	*L. casei*	Amp × 7 d	Cured/died
7	57/M	Bacteremia	Diabetes mellitus, AML, CRF, broad-spectrum ABT	*Lactobacillus* species	Amp × 4 d	?/Died
8	67/M	Intraabdominal abscess*	Alcohol abuse, CAD, asthma, liver cirrhosis, GI bleeding, broad-spectrum ABT	*L. casei*, *Bacteroides fragilis*	Vm × 2 d, Chl × 2 d	?/Died
9	26/F	Bacteremia*	Factitious disease (injecting saliva iv)	*Lactobacillus* species, *Alcaligenes dentrificans*, *Klebsiella oxytoca*, *Stenotrophomonas maltophilia*	Amp × 24 d	Cured/lived
10	69/F	Intraabdominal abscesses*	CVA, seizure	*Lactobacillus* species, *Bifidobacterium* species	Vm × 14 d, Amp × 5 d	Cured/died
11	48/F	Bacteremia*	Chronic active hepatitis, upper GI bleeding (endoscopic vasopressin)	*L. casei*, viridans group *Streptococcus*, *Candida albicans*	Cm × 2 d	?/Died
12	66/M	Endocarditis	Valvular disease, dental abscess, CVA	*L. casei*	Pen × 23 d, Tm × 5 d	Cured/died
13	65/M	Bacteremia	Esophageal cancer, TPN, broad-spectrum ABT	*Lactobacillus acidophilus*	Vm × 2 d, Pen × 1 d	?/Died
14	52/M	Bacteremia*	Laryngeal cancer (recent neck surgery), liver cirrhosis, chronic obstructive lung disease	*Lactobacillus* species, viridans group *Streptococcus*	Naf × 11 d	Cured/died
15	67/F	Bacteremia	Diabetes mellitus, liver cirrhosis, alcohol abuse, TPN	*Lactobacillus* species	Amp × 7 d, Amp/CA × 7 d	Cured/lived
16	74/M	Bacteremia	Pancreatitis, alcohol abuse, pyelonephritis, pneumonia	*Lactobacillus* species	?	?/Died
17	65/M	Bacteremia*	Renal cell cancer, radiation therapy, broad-spectrum ABT	*Lactobacillus* species, *Escherichia coli*, *Enterococcus* species	Tic/CA × 10 d	?/Lost to follow-up

Figure 6. Continued on next page.

Case no., classification	Patient's age (y)/sex	Diagnosis	Underlying condition(s) (treatment)	Organism(s) isolated	Therapy	Outcome (microbiological/clinical)
34	47/F	Bacteremia*	Marfan's syndrome, AVR, pneumonia	*Lactobacillus* species, viridans group *Streptococcus*	Vm × 7 d, Amp/CA × 7 d	Cured/lived
35	84/M	Bacteremia*	Diabetes mellitus, prostate cancer, CLL, CHF	*Lactobacillus* species, *E. coli*	Czox × 9 d	Cured/died
36	45/F	Bacteremia*	Diabetes mellitus, pancreatitis	*Lactobacillus* species, viridans group *Streptococcus*, *Anaerobic* gram-negative bacilli	Amp × 11 d	Cured/lived
37	36/M	Bacteremia*	Lung transplant (P/A/C), diabetes mellitus, broad-spectrum ABT	*Lactobacillus* species, *Alcaligenes* species, *Enterococcus* species	Imi × 34 d	Cured/died
38	66/M	Bacteremia*	Carcinoid, recent endoscopy	*Lactobacillus* species, *Veillonella parvula*	Cpfx × 34 d	Cured/died
39	66/M	Bacteremia*	Rheumatic heart disease, tricuspid valve repair	*Lactobacillus* species, viridans group *Streptococcus*	Cm × 10 d	Cured/died
Possible						
40	81/M	Bacteremia	Metastatic cancer	*Lactobacillus leichmanii*	Tic/CA × 1 d	Cured/died
41	79/F	Bacteremia	Occult metastatic adenocarcinoma, transrectal and transvaginal biopsy	*Lactobacillus* species	Ctax × 10 d	Cured/died
42	74/F	Bacteremia	Diabetes mellitus, endometrial cancer	*Lactobacillus* species	Pip × 1 d, Amp × 13 d	Cured/lived
43	62/M	Bacteremia	Esophageal carcinoma, alcohol abuse, pulmonary embolus	*Lactobacillus* species	Naf × 4 d, Cm × 6 d	Cured/died
44	64/F	Bacteremia	Breast cancer, post-chemotherapy	*Lactobacillus* species	Pen × 6 d	Cured/lived
45	46/F	Bacteremia	Urinary tract infection, Marfan's syndrome, abdominal aneurysm repair, broad-spectrum ABT	*Lactobacillus* species	Amp, Pip × 16 d	Cured/died

Case no., classification	Patient's age (y)/sex	Diagnosis	Underlying condition(s) (treatment)	Organism(s) isolated	Therapy	Outcome (microbiological/clinical)
18	59/M	Intraabdominal abscess	Rectosigmoid cancer, TPN, CRF, broad-spectrum ABT	*L. casei*	Imi × 17 d	Cured/died
19	52/F	Endocarditis	CVA	*L. casei*	Pen × 45 d	Cured/died
20	81/F	Perinephric abscess*	Diabetes mellitus, restrictive lung disease, CRF	*Lactobacillus* species, *Candida albicans*	Vm, Cfox × 5 d	Cured/died
21	74/M	Bacteremia*	Gastric cancer (recent partial gastrectomy)	*Lactobacillus* species, *E. coli*	Tic/CA × 5 d	Cured/died
22	57/M	Endovascular infection*	CAD/CABG, pancreatitis (exploratory laparotomy), broad-spectrum ABT, TPN	*L. casei*, *C. albicans*	Pen × 11 d	Cured/died
23	59/M	Endovascular infection*	Diffuse atherosclerosis, abdominal aortic aneurysm repair, ischemic lower extremity, urinary tract infection, enterovascular fistula	*Lactobacillus* species, *Enterococcus* species	Tic/CA × 16 d	Cured/died

Probable						
24	42/M	Bacteremia*	Knee surgery for streptococcal septic arthritis	*L. casei*, *C. albicans*, viridans group *Streptococcus*	Vm × 7 d, Em × 7 d	Cured/lived
25	37/F	Bacteremia*	Cardiac transplant (P/A/C), diabetes mellitus, broad-spectrum ABT	*L. casei*, *Clostridium subterminale*, *B. fragilis*	Vm × 6 d	Cured/lived
26	35/M	Bacteremia*	NHL (prednisone), CRF, GI bleeding, broad-spectrum ABT	*Lactobacillus* species, *Pediococcus* species	Cpfx × 2 d	?/Died
27	52/M	Bacteremia*	Rectal carcinoma, broad-spectrum ABT	*Lactobacillus* species, viridans group *Streptococcus*	Cman × 10 d	Cured/lost to follow-up
28	71/F	Bacteremia*	Broad-spectrum ABT, ischemic bowel, IHSS	*Lactobacillus* species, *C. albicans*, *Pseudomonas aeruginosa*	None	?/Died
29	79/F	Bacteremia*	Diabetes mellitus, TPN, diverticulosis, broad-spectrum ABT	*Lactobacillus* species, anaerobic gram-negative bacilli	Ctax × 14 d	Cured/lived
30	69/F	Bacteremia*	Sudden cardiopulmonary arrest, broad-spectrum ABT, diabetes mellitus, Alzheimer's disease, urinary tract infection	*Lactobacillus* species, *C. albicans*	Cm × 8 d, Pen × 1 d	Cured/lived
31	73/M	Bacteremia*	Prostate cancer, diabetes mellitus, sickle cell trait, CAD	*Lactobacillus* species, *C. albicans*	Vm × 6 d, Amp × 3 d; Tic/CA × 2 d	Cured/died
32	71/F	Bacteremia*	Pneumonia, Alzheimer's disease	*Lactobacillus* species, non-viridans-group *Streptococcus*	Tic/CA × 15 d	Cured/lived
33	70/M	Bacteremia*	Bladder cancer, stenosis of ilial conduit	*Lactobacillus* species, *Streptococcus bovis*, *Propionibacterium acnes*	Cman × 7 d	Cured/lived

Figure 6. Data from cases of *Lactobacillus* bacteremia. ABT = antibiotic therapy; AML = acute myelocytic leukemia; Amp = ampicillin; AVR = aortic valve replacement; CA = clavulanic acid; CABG = coronary artery bypass graft; CAD = coronary artery disease; cfox = cefoxitin; CHF = congestive heart failure; Chl = chloramphenicol; CLL = chronic lymphocytic leukemia; Cm = clindamycin; Cman = cefamandole; Cpfx = ciprofloxacin; CRF = chronic renal failure; Ctax = cefotaxime; CVA = cerebrovascular accident; Czox = ceftizoxime; DVT = deep-vein thrombosis; Em = erythromycin; GI = gastrointestinal; Gm = gentamicin; IHSS = idiopathic hypertrophic subaortic stenosis; Imi = imipenem; Naf = nafcillin; NHL = non-Hodgkin's lymphoma; P/A/C = prednisone/azathioprine/cyclosporine; Pen = penicillin; Pip = piperacillin; SLE = systemic lupus erythematosus; Tic = ticarcillin; Tm = tobramycin; TPN = total parenteral nutrition; Vm = vancomycin. * Polymicrobial bacteremia [39].

Bacteremia was polymicrobial in 27 patients and developed during hospitalization in 39. Thirty-one patients died, but only one death was attributable to *Lactobacillus* bacteremia. Lactobacilli are relatively avirulent pathogens that produce bacteremia in patients with serious underlying illnesses, many of whom have received prior antibiotic therapy that may select out for the organism. While rarely fatal in itself, *Lactobacillus* bacteremia identifies patients with serious and rapidly fatal illness.

Majamaa et al., (1997) evaluated, as first step of their study, the clinical and immunologic effects of cow's milk elimination without (n = 14) and with (n = 13) the addition of *Lactobacillus* GG (5 x 10^8 CFU/g formula) in an extensively hydrolyzed whey formula in infants with atopic eczema and cow's milk allergy [40]. The second part of the study involved 10 breast-fed infants who had atopic eczema and cow's milk allergy. In this group *Lactobacillus* GG was given to nursing mothers. The concentrations of fecal α1- antitrypsin, TNF-α, and eosinophil cationic protein were determined as markers of intestinal inflammation before and after dietary intervention. The clinical score of atopic dermatitis improved significantly during the 1-month study period in infants treated with the extensively hydrolyzed whey formula fortified with *Lactobacillus* GG. The concentration of α1-antitrypsin decreased significantly in this group ($p = 0.03$) but not in the group receiving the whey formula without *Lactobacillus* GG ($p = 0.68$).

In parallel, the median concentration of fecal TNF-α decreased significantly in this group, from 709 pg/gm (91 to 1131 pg/gm) to 34 pg/gm (19 to 103 pg/gm) ($p = 0.003$), but not in those receiving the extensively hydrolyzed whey formula only ($p = 0.38$). The concentration of fecal eosinophil cationic protein remained unaltered during therapy. These results suggested that probiotic bacteria could promote endogenous barrier mechanisms in patients with atopic dermatitis and food allergy, and by alleviating intestinal inflammation, could act as a useful tool in the treatment of food allergy.

Schiffrin et al., (1997) observed, on 28 healthy adults human volunteers (mean age: 36.9 years) who received two LAB strains, an enhanced phagocytic activity towards *Escherichia coli* [41]. The adults were randomly distributed into 2 groups, for the first 3 weeks of the study both groups received milk, in the following 3 weeks the volunteers received, daily, a fermented milk supplemented with 1 x 10^{10} CFU of *Bifidobacterium bifidum* strain Bb12 (group 1) or 7 x 10^{10} CFU of *Lactobacillus acidophilus* strain La1 (group 2) (figure 7). In the final 6 weeks of the study, both groups received milk again, but no fermented products. All volunteers consumed 120 mL milk or fermented milk 3 times a day.

Solis Pereyra et al., (1997) evaluated in mice and humans the effects of LAB on production of INF- α, β, and γ; interleukin 1β; and TNF-α in order to study the role of food in the stimulation of cytokine production. Yogurt bacteria induced plasma INF-α and β production in mice [42]. Yogurt intake containing 10^{11} bacteria led to increased 2'-5'A synthetase activity in human blood mononuclear cells.

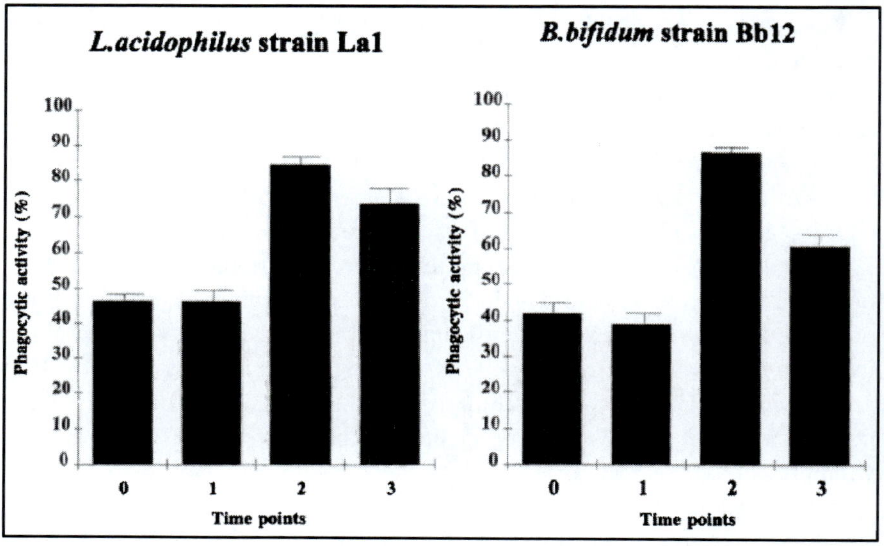

Figure 7. Percentage of phagocytes (granulocytes and monocytes) showing phagocytic activity in the peripheral blood of volunteers receiving fermented milk supplemented with *Bifidobacterium bifidum* strain Bb12 or *Lactobacillus acidophilus* strain Lal. Mean (± SEM) phagocytic activity is shown at the beginning of the study (time point 0), after subjects received milk for 3 wks (time point I) after subjects received fermented milk for 3 wks (time point 2), and 6 wks after the end of fermented milk consumption (time point 3) [41].

This result may suggest a peripheral interferon. This effect was also found when subjects consumed 10^8 bacteria/d yogurt for 15 days. In an *in vitro* model, blood mononuclear cells cultured in the presence of yogurt bacteria produced interleukin lβ, TNF, and INF-α and γ. These results suggested the involvement of a certain type of food in cytokine production under healthy conditions (figure 8).

In 1997, Marteau et al. assessed the effects of *Lactobacillus acidophilus* strain La1 on the jejunal secretions and serum concentrations of total and specific immunoglobulins and proteins [43]. Twelve healthy volunteers ingested a fermented milk containing La1 or a control from day 1 till day 28, following a randomized double blind protocol. At days 0 and 28, the jejunum was successively perfused with a control solution and with a La1 suspension. Then, serum concentrations and jejunal secretions of albumin, orosomucoid, transferrin, α2-macroglobulin, m-IgA, p-IgA, IgG, IgM, secretory component, and specific antibodies against La1 were assessed.

Food components	In vivo	In vitro	Subjects	Authors
β-Carotene	↓ Cancer risk, ↑ T helper cells, ↑ T cells expressing IL-2 receptor	↑ IL-1 and TNF production by BMCs	Humans	Abdel Fatth et al. 1993
β-Carotene	↑ CD16+ cells (NK activity)	—	Humans	Watson et al. 1991, Prabhala et al. 1990
Vitamin E	IL-1 and IL-2 induction	—	Mice	Wu et al. 1994
Lactobacillus acidophilus and *Streptococcus thermophilus*	↑ Phagocytosis	↑ Phagocytosis, ↑ IgM-producing cells against SRBCs	Mice	Perdigon et al. 1987
Lactobacillus bulgaricus and *S. thermophilus*	↓ *Salmonella enteritidis* infection	—	Mice	De Simone et al. 1988
L. bulgaricus	IFN-α and -β induction	—	Mice	Solis Pereyra et al. 1991
L. bulgaricus + *S. thermophilus*	↑ 2'-5' A synthetase	IL-1β, TNF-α, and IFN-γ induction	Humans	Solis Pereyra et al. 1993
Lactobacillus casei	↑ IgM secreting cells to rotavirus vaccine	—	Humans	Isolauri et al. 1995

Figure 8. Effect of some foods on immune response [42].

They observed that serum concentrations of IgA slightly increased between d_0 and d_{28} in the group receiving La1 (1.85 ± 0.64 *vs* 1.76 ± 0.76 g/L; $p = 0.02$) while the other parameters were not altered.

The role of intestinal bacterial flora in oral tolerance induction to the IgE response using germfree (CF) mice was interestingly evaluated by Sudo et al. in 1997 [44]. When CF mice were orally administered 20 mg of OVA as tolerogen before a systemic challenge with OVA, the T_Hl-mediated responses, such as the production of lgG2a and IFN-7, were abrogated, while the T_H2-mediated immune responses, such as the production of IgE, IgCl, and 11-4, were maintained.

Moreover, the basal level of IL-4 production *in vitro* was significantly higher in the GF mice than that of IL-4 in specific pathogen-free mice when challenged systemically with OVA. On the other hand, both T_Hl and T_H2 responses were fully sensitive to such tolerance induction in specific pathogen-free mice. The reconstitution of intestinal flora of CF mice with *Bifidobacferium infantis*, one of the predominant bacteria in the intestinal flora, restored the susceptibility of these T_H2 responses to oral tolerance induction; however, this was only effective when such reconstitution was performed in neonates, but not in mice at an older age.

These results thus suggested that intestinal bacterial flora play a crucial role in generating a T_H2 cell population whose size and response are adequately regulated and, consequently, fully susceptible to oral tolerance induction, probably by affecting the development of gut-associated lymphoid tissue at the neonatal stage.

In 1998, Maassen et al. administered some lactobacilli to mice inducing a cytokine production effective in inducing and improving the humoral response. Lactobacilli administered in SJL mice were able to enhance or inhibit the development of induced autoimmune encephalomyelitis (EAE). Such modulation was strain dependent and was confirmed by the immuno-histochemical analysis of cytokine profiles. Also in BALB/c mice, immunized with chicken gamma globulin, the serum antibody response was modulated by lactobacilli [45].

In 1998, Miettinen et al. stimulated human peripheral blood mononuclear cells (PBMC) with three nonpathogenic *Lactobacillus* strains (*Lactobacillus rhamnosus* E509, *Lactobacillus rhamnosus* E522, *Lactobacillus bulgaricus* E585) and with one pathogenic *Streptococcus pyogenes* strain, and analyzed cytokine gene expression and protein production [46]. All bacteria strongly induced IL-1β, IL-6, and TNF-α mRNA expression and protein production. *Streptococcus pyogenes* was the most potent inducer of secretion of IL-12 and

IFN-γ, and two of three *Lactobacillus* strains induced IL-12 and IFN-γ production. All strains induced IL-18 protein production. IL-10 and IL-4 production was induced weakly or not at all, respectively. Their data showed that nonpathogenic lactobacilli and pathogenic streptococci induced Th1 type cytokines IL-12, IL-18, and IFN-γ in human PBMC.

Donnet-Hughes et al., (1998) undertook a study on healthy adult volunteers (27 females and 15 males) from 21 to 57 years in order to determine the effect of a fermented milk product containing *Lactobacillus johnsonii* La1 on the phagocytic activity of peripheral blood leukocytes [47]. Volunteers were randomly distributed in three groups and all of them received 150 mL of fermented milk each day for 3 weeks: group A received a freshly prepared product fermented by *Streptococcus thermophilus*; group B received *Streptococcus thermophilus* and 10^7 CFU/mL *Lactobacillus johnsonii* La1; group C received a product stored for a period of 21 to 28 d containing *Streptococcus thermophilus* and 10^6 CFU/mL of *Lactobacillus johnsonii* La1. A survival of *Lactobacillus johnsonii* La1 to intestinal transit was observed in groups B and C whereas only in group B there was an enhancement on respiratory burst and phagocytic activity (figure 9).

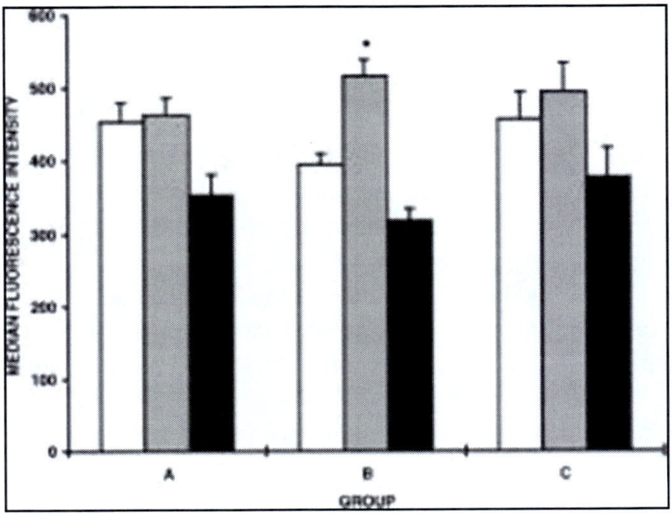

Figure 9. Median fluorescence values for phagocytic activity in the peripheral blood of volunteers receiving 0 (A), 10^7 (B), or 10^6 CFU (C) of *Lactobacillus johnsonii* La1/d at end of the equilibration period (white shading); immediately after ingestion of the different products (grey shading); 6 wks after the end of fermented milk consumption (dark shading) [47].

Spanhaak et al., (1998) evaluated the effect of consumption of milk fermented by *Lactobacillus casei* strain Shirota on the composition and metabolic activities of the intestinal microflora, and immune parameters in humans [48]. 20 healthy male subjects aged 40-65 years were selected. A placebo-controlled trial was performed with 10 subjects randomly assigned to a control and 10 to a treatment group. During the first and last two weeks of the 8-week study the subjects received a strictly controlled diet without fermented products. The same controlled diet was given during the intermediate 4 week test period but then the treatment group received three times daily 100 ml of fermented milk containing 10^9 CFU *Lactobacillus casei* Shirota/mL, whereas the same amount of unfermented milk was given to the subjects in the control group. In comparison to the control group, the consumption of *Lactobacillus casei* Shirota-fermented milk resulted in an increase of the *Lactobacillus* count in the faeces in which the administered *Lactobacillus casei* Shirota was predominant at the level of 10^7 CFU/g wet faeces. This was associated with a significant increase in *Bifidobacterium* counts ($p < 0.05$). Some shifts in the other bacterial species were found, such as a decreased number of *Clostridium*; however the differences were not statistically different between the treatment and the control groups. The β-glucuronidase and β-glucosidase activities per 10^{10} bacteria decreased significantly ($p < 0.05$) at the second week of the 4 week test period with the consumption of *Lactobacillus casei* Shirota-fermented milk. Furthermore, the consumption of the fermented milk product resulted in a slight but significant increase in the moisture content of the faecal samples ($p < 0.05$). No treatment effects were observed for any of the immune parameters measured (including natural killer (NK) cell activity, phagocytosis and cytokine production). The results suggested that consumption of *Lactobacillus casei* Shirota-fermented milk was able to modulate the composition and metabolic activity of the intestinal flora and indicated that *Lactobacillus casei* Shirota-fermented milk did not influence the immune system of healthy immunocompetent males.

In a double-blind, cross-over study, challenges with milk in milk hypersensitive and healthy adults with or without an intestinal bacterial strain (*Lactobacillus* GG ATCC 53103) were performed to test wether probiotic bacteria could reduce hypersensitivity by influencing phagocytes [49]. The challenge-induced immunoinflammatory response was recorded by measuring the expression of phagocytosis receptors prior to and after the challenge using flow cytometry. In milk-hypersensitive subjects, milk challenge increased significantly the expression of CR1, Fcγ RI and Fcα R in neutrophils and CR1, CR3 and Fcα R in monocytes. Milk with *Lactobacillus* GG prevented the increase of the receptor expression. In healthy subjects, milk challenge did not influence receptor expression while milk with *Lactobacillus* GG increased significantly the expression of CR1, CR3, Fcγ RIII and Fcα R in neutrophils (figure 10). Probiotic bacteria appeared to modulate the nonspecific immune response differently in healthy and hypersensitive subjects. This was seen as an immunostimulatory effect in healthy subjects, and as a down-regulation of immunoinflammatory response in milk-hypersensitive subjects.

Fukushima et al., (1998) performed a study to elucidate the influence of a probiotic formula on intestinal microflora and local immunity in healthy children [50]. A follow-up formula containing viable bifidobacteria was given to seven healthy Japanese children (15 to 31 months old) for 21 days (figure 11 A). During intake of the formula, the administered strain was detected in feces from five subjects (71%) and total fecal bifidobacteria slightly increased. Fecal levels of total IgA and anti-poliovirus IgA during intake of the formula were significantly higher than those before intake ($p < 0.05$). The increase in local IgA levels resulting from ingestion of the probiotic formula may contribute to enhancement of the mucosal resistance against gastrointestinal infections (Figure 11 B).

Rautio et al., (1999) described the case of a 74-year-old woman with a several-year history of hypertension and non-insulin-dependent diabetes mellitus that was admitted to because of a 2-week history of mild abdominal discomfort and a few-day history of fever(51). By means of Polymerase Chain Reaction (PCR) assay with species-specific primers and a Pulsed-Field Gel Electrophoresis (PFGE) it was found that the cause was the probiotic strain *Lactobacillus rhamnosus* GG, commonly used in dairy products in Finland since 1990.

In 1999, Yoshimura et al. hypothesized that, because of activation of cells via CD14 and Toll-like receptor (TLR) 2 and TLR4 by the LPS component of the Gram-negative bacterial cell wall, Gram-positive bacteria could also be recognized by TLRs [52]. Heterologous expression of human TLR2, but not TLR4, in fibroblasts conferred responsiveness to *Staphylococcus aureus* and *Streptococcus pneumoniae* as evidenced by inducible translocation of NF-kB. CD14 co-expression synergistically enhanced TLR2-mediated activation. To determine which components of Gram-positive cell walls activate Toll proteins, a soluble preparation of peptidoglycan prepared from *Staphylococcus aureus* was tested.

	Control subjects				Milk-hypersensitive subjects			
	Without LGG		With LGG		Without LGG		With LGG	
	Change (%)	*P*	Change (%)	*P*	Change (%)	*P*	Change (%)	*P*
CR1								
Neutrophils	104	0.72	128	0.004	122	0.02	105	0.74
Monocytes	101	0.86	108	0.15	117	0.008	96	0.38
CR3								
Neutrophils	95	0.86	133	0.05	127	0.06	122	0.15
Monocytes	83	0.14	109	0.14	137	0.02	104	0.74
FcγRI								
Neutrophils	104	0.81	109	0.07	114	0.02	104	0.64
Monocytes	105	0.86	101	0.95	100	0.74	103	0.74
FcγRII								
Neutrophils	106	0.37	110	0.09	107	0.27	97	0.31
Monocytes	99	0.81	106	0.14	98	0.84	105	0.08
FcγRIII								
Neutrophils	104	0.21	108	0.04	100	0.95	104	0.15
Monocytes	114	0.31	91	0.37	177	0.15	117	0.41
IgαR								
Neutrophils	102	0.68	110	0.03	114	0.02	102	0.95
Monocytes	96	0.68	104	0.14	113	0.008	93	0.20

Figure 10. Receptor expression changes as percentage values during 1 week milk challenge (after/prior to the challenge x 100%) in control and in milk-hypersensitive subjects with milk only (without LGG) or with milk containing *Lactobacillus* GG (with LGG). No changes during the challenge = 100% [49].

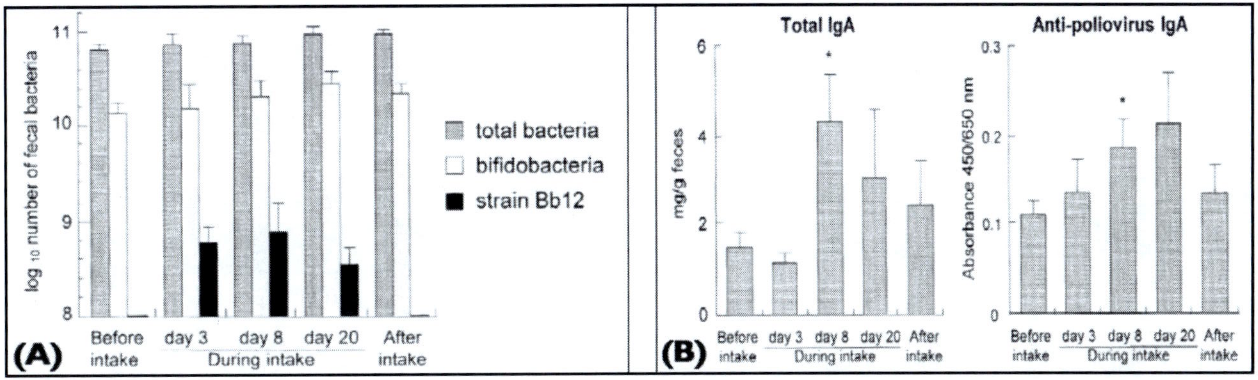

Figure 11. (A) Effect of probiotic formula for children on fecal microflora. Data are expressed as mean ± SEM. The total number of fecal bacteria is the summation of the numbers of bacteroidaceae, bifidobacteria, eubacteria, anaerobic gram-positive cocci, clostridia, enterobacteriaceae, streptococci, and staphylococci. The number of children was seven except on day 3 and day 20 ($n = 6$), because fecal samples from two subjects were not obtained; (B) Effect of probiotic formula for children on fecal IgA. Data are expressed as mean ± SEM. Statistical differences were calculated by Mann–Whitney test by comparison with the value before intake (*$p < 0.05$)[50].

To examine the potential role of TLRs in responses to bacteria, clonal cell lines were engineered in which epitope-tagged chimeric proteins were expressed at high levels. CHO cells did not express a functional TLR2 transcript, in contrast they expressed a functional mRNA encoding TLR4. They exposed CHO transfectants to heat-killed *Staphylococcus aureus* for 45 min and subsequently subjected nuclear extracts from these cells to the EMSA to assess for the presence of induced NF-kB. No nuclear translocation of NF-kB was observed but CHO/TLR2 cells were activated with heat-killed *Staphylococcus aureus* at the highest concentration tested.

Next, they tested the responses of the CHO/TLR cell lines using *Streptococcus pneumoniae*. Unlike heat-killed Staphylococcus aureus, 10^8 CFU of heat-killed *Streptococcus pneumoniae* partially stimulated CHO/CD14 cells in the absence of TLR over expression. Despite the efforts, they were unable to observe the direct binding of bacteria to TLRs and a prominent possibility could be that Ag exposure results in the processing of an endogenous ligand that activates TLRs. A possible explanation of how TLRs function provide specificity of recognition for diverse types of bacteria could be the presence of a common family of signaling receptors by seemingly diverse bacteria could explain why the clinical picture of sepsis caused by Gram-negative bacteria and Gram-positive bacteria.

In 1999, Madsen et al. tried to determine the role of colonic aerobic luminal bacteria and *Lactobacillus* species in the development of colitis in interleukin (IL)-10 gene-deficient mice [53]. Intestine from 2-16-week-old mice was scored histologically and cultured for bacteria. *Lactobacillus* spp. repopulation of the colonic lumen was achieved via daily rectal delivery of *Lactobacillus reuteri* or oral lactulose therapy. At 2 weeks of age, IL-10 gene-deficient mice showed no colonic injury but did display abnormal colonic bacterial colonization with increased colonic mucosal aerobic adherent and translocated bacteria in conjunction with reduced *Lactobacillus* spp. levels. In association with the abnormal colonic bacterial colonization, colitis developed by 4 weeks of age. Restoring *Lactobacillus* spp. to normal levels reduced levels of colonic mucosal adherent and translocated bacteria and attenuated the development of the colitis. In the neonatal period, IL-10 gene-deficient mice have decreased levels of colonic *Lactobacillus* spp. and an increase in colonic mucosal adherent and translocated bacteria. Normalizing *Lactobacillus* spp. levels reduced colonic mucosal adherent and translocated bacteria and prevented colitis.

Still in 1999, Perdigón et al. studied IgA^+ and IgM^+ B cells comparatively in bronchus and intestine and $CD4^+$ T cells and IgA anti-LAB antibodies in the intestinal fluid, induced by oral administration of *Lactobacillus casei* CRL 431, *Lactobacillus delbrueckii* ssp. *bulgaricus* CRL 423, *Lactobacillus acidophilus* CRL 924, *Lactobacillus plantarum* CRL 936, *Lactobacillus rhamnosus* CRL 74, *Lactococcus lactis* CRL 526 and *Streptococcus salivarius* ssp. *thermophilus* CRL 412 [54]. The increase in the IgA^+ B cells in the bronchus meant that these LAB were able to induce the IgA cycle by interaction with M cells from Peyer's patches or intestinal epithelial cells (figure 12).

The IgM^+ cells increased when the stimulus did not induce the switch from IgM^+ to IgA^+. The increase in the $CD4^+$ cells suggested an interaction of Peyer's patches and enhancement of the B- and T-cell migration. *Lactobacillus casei* and *Lactobacillus plantarum* interacted with Peyer's patch cells showing an increase in IgA^+, $CD4^+$ cells and antibodies specific for the stimulating strain. *Lactobacillus acidophilus* induced gut mucosal activation by interaction with the epithelial cells without increase in the immune cells associated with the bronchus. Although *Lactobacillus rhamnosus* and *Streptococcus salivarius* subsp. *thermophilus* interacted with epithelial cells, they also induced an immune response against their epitopes.

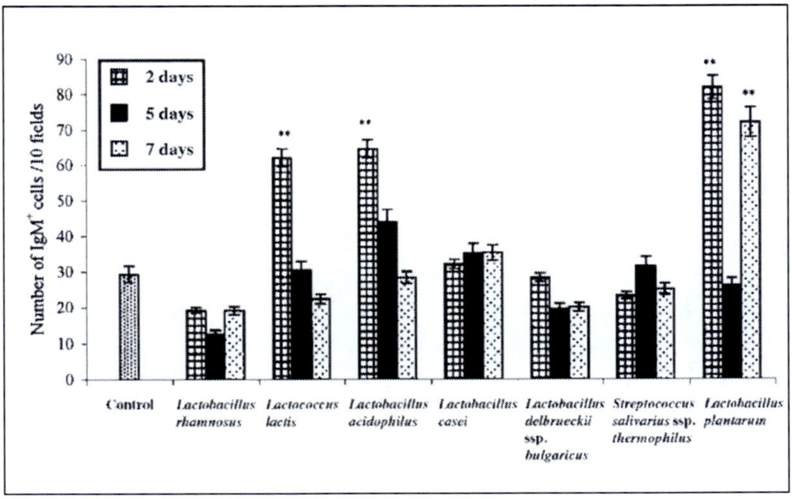

Figure 12. Effect of oral administration of different LAB on the number of IgM cells present in lamina propria of the small intestine tissue. The IgM B cells were measured by an immunofluorescence test using μ-chain monospecific antibody after 2, 5, or 7 d of feeding with each LAB assayed. The bars indicate arithmetic means (± SD) of the number of cells in 10 fields (100 X) expressing the marker [54].

Lactococcus lactis and *Lactobacillus delbrueckii* subsp. *bulgaricus* induced an increase of IgA^+ cells entering the IgA cycle but not $CD4^+$ cells; thus, these bacteria could have been bound to epithelial cells that activated B lymphocytes without processing and presenting of their epitopes.

Herãas et al., in 1999, studied the effect of the probiotic strain *Lactobacillus plantarum* 299v on the immune functions of gnotobiotic rats [55]. One group of germ-free rats was colonized with the type 1-fimbriated *Escherichia coli* O6:K13:H1 and another group with the same *Escherichia coli* strain together with *Lactobacillus* plantarum 299v. One and 5 weeks after colonization, bacterial numbers were determined in the contents of the small intestine, caecum and mesenteric lymph nodes (figure 13).

Rats given *Lactobacillus plantarum* in addition to *Escherichia coli* showed lower counts of *Escherichia coli* in the small intestine and in caecum 1week after colonization, compared with the group colonized with *Escherichia coli* alone, but similar levels after 5 weeks. Rats colonized with *Lactobacillus plantarum* + *Escherichia coli* had significantly higher total serum IgA levels and marginally higher IgM and IgA antibody levels against *Escherichia coli* than those colonized with *Escherichia coli* alone. They also showed a significantly increased density of $CD25^+$ cells in the lamina propria and displayed a decreased proliferative spleen cell response after stimulation with concanavalin A or *Escherichia coli* 1 week after colonization.

Still in 1999, Kato et al. examined whether *Lactobacillus casei* strain Shirota (LcS) could induce the production of interleukin IL-12 and IFN-γ from murine splenocytes *in vitro* [56].

Stimulation by LcS induced a marked production of IL-12 by X-ray-irradiated splenocytes (X-irr-Spl). The production of IL-12 by X-irr-Spl was independent of the presence of nylon wool column passed splenocytes (NW-Spl) (figure 14).

IFN-γ was produced by splenocytes by the stimulation with concanavalin A (Con A). LcS showed a synergistic stimulatory effect on the ConA-induced production of IFN-γ, while the IFN-γ production was reduced by anti-IL-12 antibody treatment. NW-Spl produced IFN-γ following treatment with recombinant IL-12. IFN-γ production by splenocytes was the result of the production of IL-12 from X-irr-Spl stimulated by LcS. Furthermore in BALB/c mice the oral administration of viable LcS augmented the production of IFN-γ but not that of IL-4 or IL-5 by splenocytes (figure 15).

It was suggested that LcS could primarily activate X-irr-Spl, probably macrophages, and these cells secreted IL-12. The IL-12 induced by LcS stimulated the production of IFN-γ.

	Lumenal population levels (CFU/g contents)			
	Small intestine		Caecum	
	1 week	5 weeks	1 week	5 weeks
E. coli colonization				
E. coli	$10^{5.9 \pm 0.4}$	$10^{8.5 \pm 0.7}$	$10^{9.8 \pm 0.1}$	$10^{10.2 \pm 0.2}$
E. coli + L. *plantarum* colonization				
E. coli	$< 10^{5}$**	$10^{8.3 \pm 0.9}$	$10^{9.3 \pm 0.2}$*	$10^{9.8 \pm 0.1}$
L. plantarum	$< 10^{5}$	$10^{7.3 \pm 2.2}$	$10^{6.6 \pm 0.4}$	$10^{8.1 \pm 1.2}$

Figure 13. Bacterial levels in the caecum and small intestinal contents [55].

LcS conc.	IL-12 concentration (pg/ml) Incubation time (h)				
(μg/ml)	1	3	6	24	48
25	n.d.[a]	n.d.	38.7	1559.1	>2500
12.5	n.d.	n.d.	23.6	1144.4	2312.3
6.25	n.d.	n.d.	18.4	1225.8	>2500
3.13	n.d.	n.d.	17.9	746.7	2054.8
0	n.d.	n.d.	n.d.	n.d.	n.d.

Figure 14. Effect of the addition of heat-killed LcS on IL-12 production by X-ray-irradiated splenocytes [56].

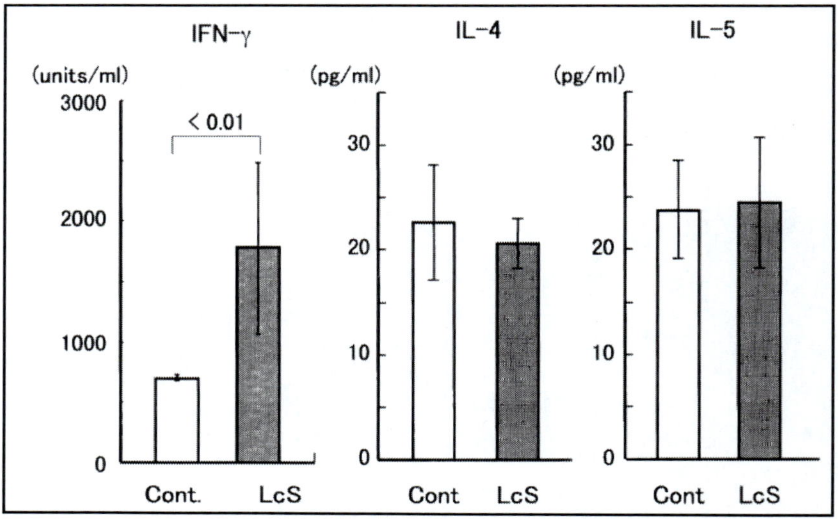

Figure 15. The oral administration of LcS augmented the production of IFN-γ but not IL-4 and IL-5 by splenocytes [56].

Takagi et al., (1999) tried to determine whether tumor induction by 3-methylcholanthrene (MC), could be inhibited by oral administration of *Lactobacillus casei* strain Shirota (LC) [57]. C3H/HeN mice were divided into four groups and assigned to the following treatments: treated with MC and given control or LC-containing diet; treated with vehicle only and given control or LC-containing diet. MC (1 mg) was injected intradermally at 7 weeks of age and the tumor incidence was monitored; LC was mixed into a diet at a concentration of 0.05% (w/w) and the diet was fed from the day of MC injection throughout the study.

Spleen cells were analyzed for the immune parameters at 12 and 16 weeks after the MC injection. Oral feeding of mice with LC reduced tumor incidence ($p < 0.05$). MC treatment lowered the *in vitro* response to concanavalin A (Con A) of spleen cells, the secretion of interleukin-2 in spleen cell culture after stimulation of the cells with Con A and the proportions of $CD3^+$ $CD4^+$ and $CD8^+$ splenic cells. However, the analysis of the spleen cells obtained from the mice treated with MC and given the LC-containing diet revealed that these disrupted host immune parameters were maintained at the level of normal controls. These results suggested that oral feeding of mice with LC inhibited MC-induced tumorigenesis by modulating the disrupted host immune responses during MC carcinogenesis.

Chapter III

The New Era of Probiotics

In 2000, Miettinen et al. demonstrated that both nonpathogenic *Lactobacillus rhamnosus* GG and pathogenic *Streptococcus pyogenes* (group A streptococci) induce NF-kB and STAT DNA-binding activity in human primary macrophages [58]. NF-kB activation was rapid and was not inhibited by a protein synthesis inhibitor cycloheximide, suggesting that these bacteria could directly activate NF-kB. STAT1, STAT3, and IFN regulatory factor-1 DNA binding was induced by both bacteria with delayed kinetics compared with NF-kB. In addition, streptococci induced the formation of IFN-α-specific transcription factor complex and IFN-stimulated gene factor-3 (ISGF3). STAT1 and STAT3 activation and ISGF3 complex formation were inhibited by cycloheximide (CHX) or by neutralization with IFN-α/β-specific Abs. Streptococci were more potent than lactobacilli in inducing STAT1, ISGF3, and IFN regulatory factor-1 DNA binding and the activation of the IFN-α signaling pathway by streptococci could play a role in the pathogenesis of these bacteria. These results indicated that extracellular Gram-positive bacteria activated transcription factors involved in cytokine signaling by two mechanisms: directly, leading to NF-kB activation, and indirectly via cytokines, leading to STAT activation.

Still in 2000, Vitini et al. studied the influence of the oral administration of different species of LAB such as *Lactobacillus casei*, *Lactobacillus acidophilus*, *Lactobacillus rhamnosus*, *Lactobacillus delbrueckii* subsp. *bulgaricus*, *Lactobacillus plantarum*, *Lactococcus lactis* and *Streptococcus thermophilus*(59). They demonstrated that all the bacteria assayed were able to increase the number of IgA producing cells associated to the lamina propria of small intestine and that the effect was dose dependent. The increase in IgA^+

producing cells was not always correlated with an increase in the $CD4^+$ T cell number, indicating that some LAB assayed only induced clonal expansion of B cells triggered to produce IgA. Most of them induced an increase in the number of cells involved in the inflammatory immune response. $CD8^+$ T cell were diminished or not affected, with exception of *Lactobacillus plantarum* that induced an increase at low dose. This fact meant that LAB were unable to induce cytotoxicity mechanisms and that LAB could be used as gut mucosal adjuvant. The different behaviors observed among them on the gut mucosal immune response, especially those that induce inflammatory immune response, showed that not all the LAB can be used as oral adjuvant and that the beneficial effect of them cannot generalized to genous or specie. The immunoadjuvant capacity could be a property of the strain assayed.

Nagao et al., (2000) administered to 9 healthy volunteers 4 x 10^{10} cells of *Lactobacillus casei* strain Shirota daily for 3 weeks and observed an increased NK activity ($p < 0.01$) [60].

Hessle et al., (2000) compared the capacities of gram-positive and gram-negative bacteria to induce IL-10 and IL-12 [61]. Monocytes from blood donors were stimulated with UV-killed bacteria from each of seven gram-positive and seven gram-negative bacterial species representing (figure 16).

Gram-positive bacteria induced much more IL-12 than did gram-negative bacteria (median, 3.500 versus 120 pg/ml at an optimal dose of 25 bacteria/cell; $p < 0.001$), whereas Gram-negative bacteria preferentially stimulated secretion of IL-10 (650 versus 200 pg/mL; $p < 0.001$). Gram-positive species also induced stronger major histocompatibility complex class II-restricted IFN-γ production in unfractionated blood mononuclear cells than did gram-negative species (12,000 versus 3.600 pg/mL; $p < 0.001$). The poor IL-12-inducing capacity of gram-negative bacteria was not remediated by addition of blocking anti-IL-10 antibodies to the cultures. No isolated bacterial component could be identified that mimicked the potent induction of IL-12 by whole gram-positive bacteria, whereas purified LPS induced IL-10. The results suggested that gram-positive bacteria induced a cytokine pattern that promoted Th1 effector functions.

Haller et al., in 2000, co-cultivated enterocyte-like CaCO-2 cells with human blood leucocytes in separate compartments of transwell cultures to address the question of whether non-pathogenic bacteria modify the immune response of the intestinal epithelium [62]. Challenge of CaCO-2 cells with non-pathogenic *Escherichia coli* and *Lactobacillus sakei* induced expression of IL-8, MCP-1, IL-1β, and TNF-α mRNA in the presence of underlying leucocytes (figure 17 A).

Bacterial species (strains)	Isolation site	Metabolism	concn (pg/ml) OF		IL-12/IL-10
			IL-12	IL-10	ratio
Gram positive					
Bifidobacterium adolescensis (18363)	Adult intestine	Anaerobic	4300±810	160±28	27
Clostridium perfrigens (1795)	Bovine	Anaerobic	160±4	120±13	1.3
Corynebacterium minutissimum (541)	Erythasma, trunk	Aerobic	2000±340	390±99	5.1
Enterococcus fecalis (19916)	Unknown	Aerobic	3500±660	200±47	18
Lactobacillus plantarum (67b)	Rectum, healthy	Aerobic	2200±540	160±50	14
Staphylococcus aureus (1800)	Pleural fluid	Aerobic	4000±850	240±83	17
Streptococcus mitis (31611)	Oral cavity	Aerobic	3600±490	200±44	18
Gram negative					
Bacteriodes vulgatus (4940)	Unknown	Anaerobic	32±2	350±39	0.09
Escherichia coli (24)	Urine, cystitits	Aerobic	140±60	1200±280	0.12
Haemophilus influenzae (21594)	Pharynx, healthy	Aerobic	90±23	560±100	0.16
Haemophilus parainfluenzae (12836)	Septic finger	Aerobic	120±50	650±120	0.18
Neisseria sicca (23929)	Pharynx, healthy	Aerobic	460±160	690±150	0.67
Pseudomonas aeruginosa (551)	Unknown	Aerobic	<30	480±120	0.06
Veillonella parvula (5123)	Intestinal tract	Anaerobic	1600±480	770±200	2.08

Figure 16. Cytokine response of human monocytes to various Gram-positive and Gram-negative bacteria. (Modified from Hessle et al., 2000) [61].

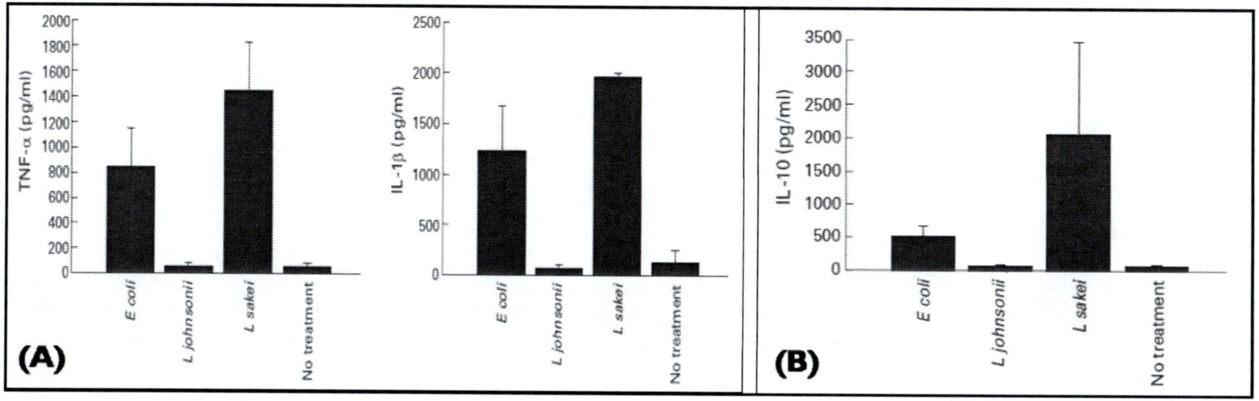

Figure 17. (A) Bacteria induced TNF-α and IL-1β response by CaCO-2/leucocyte co-cultures. Stimulation of CaCO-2/leucocyte co-cultures with non-pathogenic *Escherichia coli*, *Lactobacillus johnsonii*, and *Lactobacillus sakei* (16 hours, 10^7 CFU/mL). Secretion of TNF-α (A) and IL-1β (B) into the basolateral compartment was determined (bar charts). Culture medium (no treatment) was used as a control. Reverse transcription-polymerase chain reaction (RT-PCR) analysis was used to determine expression of TNF-α (A) and IL-1β (B) mRNA in leucocyte sensitised CaCO-2 cells; (B) induction of interleukin (IL)-10 response by CaCO-2/leucocyte co-cultures. Secretion of IL-10 wasdetermined in the basolateral compartment on stimulation of CaCO-2/leucocyte co-cultures with non-pathogenic *Escherichia coli, Lactobacillus johnsonii*, and *Lactobacillus sakei* (16 h, 10^7 CFU/mL), respectively (bar charts). Culture medium (no treatment) was used as a control. Reverse transcription-polymerase chain reaction (RT-PCR) analysis was used to determine IL-10 mRNA expression in conditioned PBMC [62].

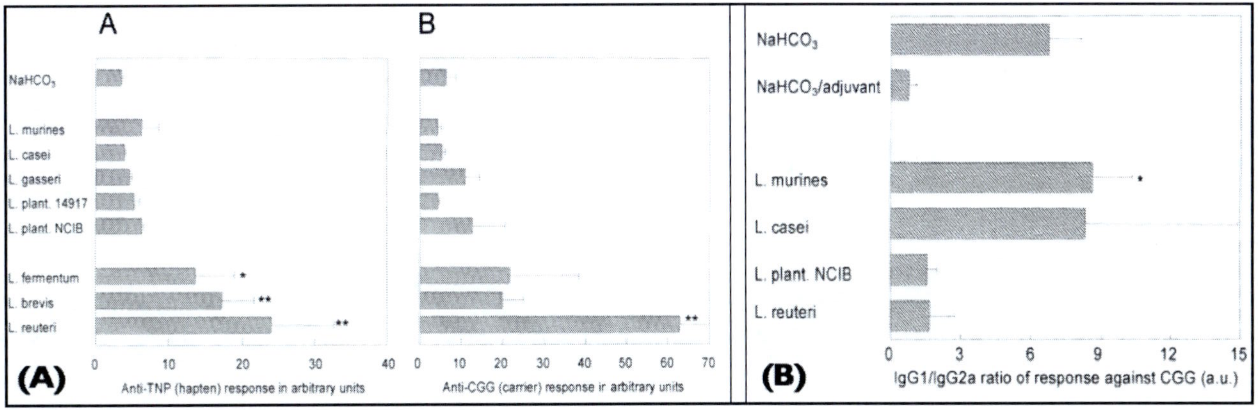

Figure 18. (A) Adjuvant activity of orally administered *Lactobacillus* strains. The effect of orally administered *Lactobacillus* strains on the antibody response against the thymus-dependent antigen TNP ± CGG administered intraperitoneally was analyzed by ELISA. The mice received a *Lactobacillus* strain or $NaHCO_3$ only orally on 4 consecutive days. On the first day the mice were also immunized i.p. with TNP ± CGG in PBS. From day 21 on this protocol was repeated. The left panel shows the mean IgG antibody response ± S.E.M. against TNP per group in arbitrary units 7 days after the second immunization. The right panel shows the IgG antibody response ± S.E.M. against the carrier protein CGG. Only significant differences compared to the $NaHCO_3$ group are indicated by asterisk; (B) IgG1/IgG2a ratio of anti-CGG response after oral administration of *Lactobacillus* strains. The effect of four orally administered *Lactobacillus* strains on the IgG1/IgG2a ratio of the response against CGG was analyzed by ELISA. The mice received a *Lactobacillus* strain or $NaHCO_3$ only orally on 4 consecutive days. On the first day the mice were also immunized i.p. with TNP±CGG in PBS. A control group was fed $NaHCO_3$ and immunized i.p. with TNP ± CGG in specol ($NaHCO_3$/adjuvant). From day 21 on, this protocol was repeated. Seven days after the second immunization the IgG1 and IgG2a responses against CGG were calculated in arbitrary units with the use of reference curves. The mean IgG1/IgG2a ratio per group2 SEM is shown [63].

Leucocyte sensitized CaCO-2 cells produced TNF-α and IL-1β whereas IL-10 was exclusively secreted by human peripheral blood mononuclear cells (figure 17 B). CaCO-2 cells alone remained hyporesponsive to the bacterial challenge. *Lactobacillus johnsonii* showed reduced potential to induce pro-inflammatory cytokines but increased TGF-β mRNA in leucocyte sensitized CaCO-2 cells. TNF-α was identified as one of the early mediators involved in cellular cross talk. In the presence of leucocytes, discriminative activation of CaCO-2 cells was observed between enteropathogenic *Escherichia coli* and non-pathogenic bacteria. The differential recognition of non-pathogenic bacteria by CaCO-2 cells required the presence of underlying leucocytes. These results strengthen the hypothesis that bacterial signaling at the mucosal surface is dependent on a network of cellular interactions.

Maassen et al., (2000) analyzed eight different common *Lactobacillus* strains with respect to mucosal induction of pro- and anti-inflammatory cytokines, IgA-producing plasma cells in the gut, as well as systemic antibody responses against a parenterally administered antigen [63]. Immunohistochemical analysis of cytokine-producing cells in the gut villi showed no significant induction of the cytokines IL-1α, IFN-γ, IL-4 or IL-10 after oral administration of wild type *Lactobacillus* strains. In contrast, oral administration of *Lactobacillus reuteri* and *Lactobacillus brevis* induced expression of the proinflammatory/T_H1 cytokines TNF-α, IL-2 and/or IL-1β. Oral administration of these two strains and *Lactobacillus fermentum* also significantly enhanced the IgG response (fig 18 A) against parenterally administered haptenated chicken gamma globulin (TNP ± CGG). The five other strains did not show this effect. *Lactobacillus reuteri* induced relatively high levels of IgG2a compared to *Lactobacillus murines*, a nonadjuving *Lactobacillus* strain (fig 18 B). Different *Lactobacillus* strains induced distinct mucosal cytokine profiles and possessed differential intrinsic adjuvanticity. Rational *Lactobacillus* strain selection could provide a strategy to influence cytokine expression and thereby influence immune responses.

Arunachalam et al., (2000) evaluated the effects of dietary consumption of *Bifidobacterium lactis* (strain HN019, DR10TM) on natural immunity of 25 healthy elderly volunteers (median age 69 y; range 60-83 y)[64]. Twelve control subjects consumed 180 mL low-fat/low-lactose milk twice daily for a period of 6 weeks; 13 test subjects consumed milk supplemented with 1.5 x 10^{11} CFU of *Bifidobacterium lactis* twice daily. Indices of natural immunity, including INF production, phagocytic capacity and phagocyte-mediated bactericidal activity, were determined via peripheral blood at 0, 3, 6 and 12 weeks post-trial commencement. Subjects fed with milk containing

Bifidobacterium lactis for 6 weeks produced significantly enhanced levels of INF-α, upon stimulation of their peripheral blood mononuclear cells in culture, in comparison to the placebo control group who received milk alone. There were also significant increases in polymorphonuclear cell phagocytic capacity among test group subjects, following consumption of milk supplemented with *Bifidobacterium lactis*, while individuals who consumed *Bifidobacterium lactis*-supplemented milk or milk alone showed enhanced phagocyte-mediated bactericidal activity. Dietary consumption of *Bifidobacterium lactis* HN019 could enhance natural immunity in healthy elderly subjects, and a relatively short-term dietary regime (6 weeks) could be sufficient to impart measurable improvements in immunity that may offer significant health benefits to consumers.

Chiang et al., (2000) tried to define the cellular basis for immune enhancement by a probiotic lactic acid bacteria strain (*Bifidobacterium lactis* HN019); and to determine whether immune enhancement could be optimized by delivery in oligosaccharide-enriched low-fat milk [65]. 50 healthy Taiwanese citizens (age range 41-81; median 60), randomly allocated to two groups, were chosen. In stage 1 (run-in control stage) all subjects consumed reconstituted low-fat milk (LFM) for 3 weeks; in stage 2 (probiotic intervention) subjects consumed *Bifidobacterium lactis* in LFM (group A) or *Bifidobacterium lactis* in lactose-hydrolysed LFM (group B) for 3 weeks; in stage 3 all subjects returned to non-supplemented LFM for a further 3 weeks (washout stage). The innate immune functions of polymorphonuclear cells (PMN) and natural killer cells (NK) were assessed at four time points via *in vitro* analyses on peripheral blood samples. While consumption of LFM alone had no significant effect on immune responses, stage 2 results indicated significantly enhanced PMN cell phagocytosis and NK cell tumor killing activity following consumption of milk containing *Bifidobacterium lactis*. These increased levels dropped following *Bifidobacterium lactis* withdrawl, but remained above the pre-treatment values. Increases in PMN and NK cell activity were greatest among subjects who consumed *Bifidobacterium lactis* in lactose-hydrolysed LFM. Thus, dietary consumption of the probiotic bacterium *Bifidobacterium lactis* HN019 enhanced immune function of two different types of leucocytes; the degree of enhancement was increased by consuming *Bifidobacterium lactis* in an oligosaccharide-rich substrate.

He et al., (2000) randomized 30 healthy volunteers into three different treatment groups consuming *Lactobacillus* GG, *Lactococcus lactis* or placebo (ethyl cellulose) for 7 days [66]. On days 1, 3 and 5, an attenuated *Salmonella typhi* Ty21a oral vaccine was given to all subjects to mimic an enteropathogenic infection. All subjects responded well to the vaccine, but no significant differences were observed in numbers of IgA-, IgG- and IgM-secreting cells among the different groups. There was a trend towards a greater increase in specific IgA among the subjects receiving the vaccine in combination with *Lactobacillus* GG. Those receiving *Lactococcus lactis* with their vaccine evinced significantly higher CR3 receptor expression on neutrophils than those receiving either the placebo or *Lactobacillus* GG (fig 19). Probiotics could influence differently the immune response to oral *Salmonella typhi* vaccine and the immunomodulatory effect of probiotics could be strain-dependent.

Receptor	LGG[b] ($n = 10$)	L ($n = 10$)	P ($n = 9$)	ANOVA P-value	Kruskal–Wallis P-value
CR1					
neutrophils	97.5 ± 24.8[a]	120.0 ± 29.9	96.0 ± 31.1	0.32	0.22
monocytes	96.4 ± 23.8	136.4 ± 59.1	118.3 ± 34.1	0.29	0.42
CR3					
neutrophils	97.8 ± 16.5	149.2 ± 41.9	112.6 ± 30.3	0.02	0.03
monocytes	101.6 ± 28.8	145.9 ± 51.5	137.8 ± 41.3	0.19	0.27
FcγRI					
neutrophils	166.0 ± 115.3	147.1 ± 129.6	107.9 ± 53.5	0.28	0.41
monocytes	108.1 ± 41.9	111.0 ± 28.6	103.8 ± 42.1	0.98	0.83
FcγRII					
neutrophils	95.0 ± 11.4	100.8 ± 7.8	98.0 ± 12.3	0.40	0.40
monocytes	101.3 ± 45.9	101.1 ± 58.7	152.0 ± 83.0	0.29	0.39
FcγRIII					
neutrophils	95.9 ± 9.1	97.9 ± 16.9	101.7 ± 16.1	0.83	0.80
monocytes	110.7 ± 54.3	118.2 ± 44.3	94.4 ± 41.8	0.71	0.62

Figure 19. Changes in FcQR and CR receptor expression of neutrophils and monocytes in subjects after oral immunization [66].

Isolauri et al., (2000) assessed the potential of probiotics to control allergic inflammation at an early age in a randomized double-blind placebo-controlled study [67]. A total of 27 infants, mean age 4.6 months, who expressed atopic eczema during exclusive breast-feeding and who had no exposure to any infant or substitute formula were weaned to probiotic-supplemented, *Bifidobacterium lactis* Bb-12 or *Lactobacillus* strain GG (ATCC 53103), extensively hydrolyzed whey formulas or to the same formula without probiotics. The extent and severity of atopic eczema, the growth and nutrition of infants, and concentrations of circulating cytokines/chemokines and soluble cell surface adhesion molecules in serum and methyl-histamine and eosinophilic protein X in urine were determined. The SCORAD score reflecting the extent and severity of atopic eczema was 16 (7-25) during breast-feeding, median (interquartile range). After 2 months, a significant improvement in skin condition occurred in patients given probiotic-supplemented formulas, as compared to the unsupplemented group; $p = 0.002$. SCORAD decreased in the *Bifidobacterium lactis* Bb-12 group to 0 (0-3.8), and in the *Lactobacillus* GG group to 1 (0.1-8.7), vs unsupplemented 13.4 (4.5-18.2), median (interquartile range), in parallel with a reduction in the concentration of soluble CD4 in serum and eosinophilic protein X in urine. The results provide the first clinical demonstration of specific probiotic strains modifying the changes related to allergic inflammation. The data further indicated that probiotics could counteract inflammatory responses beyond the intestinal milieu. The combined effects of these probiotic strains could guide infants through the weaning period, once sensitization to newly encountered antigens would have initiated.

Gill and Rutherfurd (2001) reported that oral delivery of 10^9 live and heat-killed *Lactobacillus rhamnosus* HN001 per day to mice stimulated phagocytic activities in blood and peritoneal leukocytes (figure 20), in which the latter case was dose-dependent (Figure 21) [68].

	Phagocytically active cells (%)			
	Peritoneal cells		Blood cells	
	Mean	SEM	Mean	SEM
Controls	64·5	2·6	13·7	0·5
Live HN001	75·1***	1·3	20·9***	1·1
Heat-killed HN001	71·9*	1·9	19·8**	0·5

Figure 20. Immunostimulatory effects of live or heat-killed *Lactobacillus rhamnosus* HN001 [68].

Still in 2001, Gill et al. evaluated, on 30 healthy elderly volunteers (mean age: 69 years), whether dietary supplementation with *Bifidobacterium lactis* HN019 could enhance aspects of their cellular immunity [69]. The dietary trial supplementation was divided in 3 stages lasting 9 weeks. During stage 1, subjects consumed low-fat milk (200 mL twice daily for 3 weeks) as a base-diet control. During stage 2, they consumed milk supplemented with *Bifidobacterium lactis* HN019 in a typical dose (5 x 10^{10} organisms/d) or a low dose (5 x 10^9 organisms/d) for 3 weeks. During stage 3, they consumed low-fat milk for 3 weeks. Increases in the proportions of total, helper ($CD4^+$), and activated ($CD25^+$) T lymphocytes and natural killer cells were measured in the subjects' blood after consumption of *Bifidobacterium lactis* HN019. The *ex vivo* phagocytic capacity of mononuclear and polymorphonuclear phagocytes and the anticancerogenic activity of natural killer cells were also elevated after *Bifidobacterium lactis* HN019 consumption (figure 22).

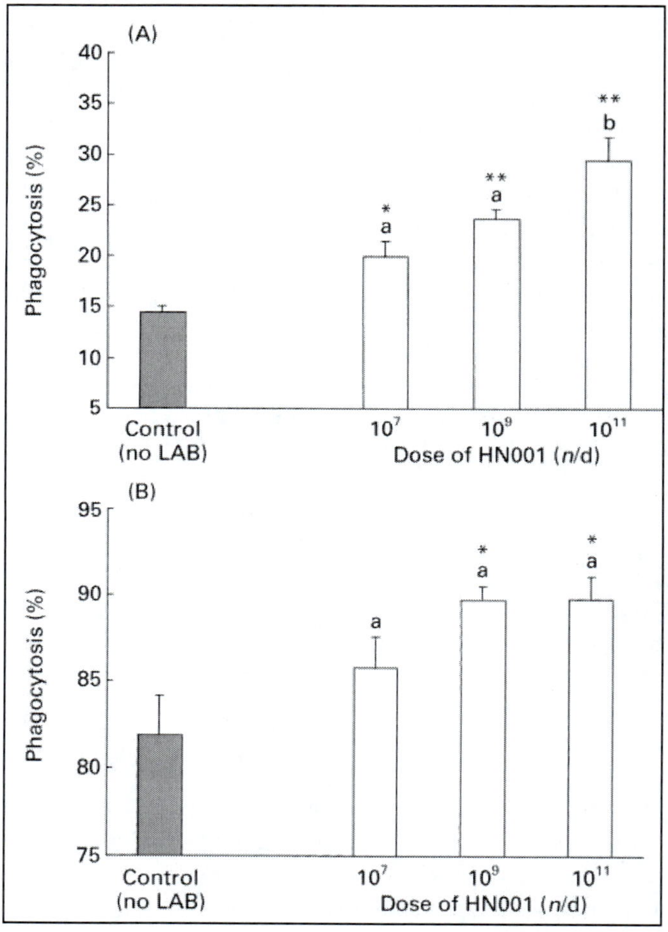

Figure 21. Dose–response effect of HN001 on enhancement of leucocyte phagocytosis. (A), blood cell phagocytosis; (B), peritoneal cell phagocytosis. Mice were fed different doses of *L. rhamnosus* HN001 by oral gavage daily for 14 d. Phagocytic activity was determined by uptake of fluoresceinated *Escherichia coli*. Values are the mean percentage of cells of each population showing phagocytic activity, with standard errors represented by vertical bars, for nine samples per group. Mean values were significantly greater than those of control mice: * $p < 0.01$; ** $p < 0.001$. ^{a,b}Mean values with unlike superscript letters were significantly different ($p < 0.05$).

	Time point			
	1 (baseline)	2 (after 3 wk consumption of unsupplemented LFM)	3 (after 3 wk supplementation with *B. lactis* HN019)[2]	4 (3 wk after cessation of *B. lactis* HN019)
		%		
T cells				
Total ($CD3^+$)	65.1 ± 2.1[3]	64.9 ± 2.6[3]	67.9 ± 2.6	65.9 ± 2.7
$CD4^+$	41.0 ± 1.6[4]	41.7 ± 1.8[4]	45.2 ± 1.3	43.7 ± 1.9
$CD8^+$	19.2 ± 2.1	20.4 ± 2.0	19.6 ± 2.1	19.5 ± 2.1
$CD25^+$	6.4 ± 0.7[4]	6.9 ± 0.7[4]	11.0 ± 1.0	7.3 ± 0.7[4]
B cells				
$CD19^+$	10.7 ± 1.4	10.3 ± 1.2	11.2 ± 1.1	9.5 ± 1.1
NK cells				
$CD56^+$	14.8 ± 1.3[4]	15.9 ± 1.4[5]	18.2 ± 1.4	16.9 ± 1.6
APCs				
HLA-DR^+	13.7 ± 1.4	15.4 ± 1.3	15.1 ± 1.3	14.5 ± 1.1

Figure 22. Overall changes in immune variables [69].

O'Mahony et al., (2001) conducted a controlled feeding trial in IL-10 knockout mice using the probiotic *Lactobacillus salivarius* ssp. *salivarius* UCC118 in order to determine the effect of probiotic consumption on the gastrointestinal microflora, tumor development and colitis in IL-10 knockout mice [70]. 20 IL-10 knockout mices were studied (10 consumed probiotic organisms in milk while 10 consumed unmodified milk) for 16 weeks. Faecal microbial analysis was performed weekly to enumerate excretion of the probiotic UCC118, total lactobacilli, *Clostridium perfringens*, bacteroides, coliforms, bifidobacteria and enterococci. At sacrifice, the small and large bowel were microbiologically and histologically assessed (figure 23).

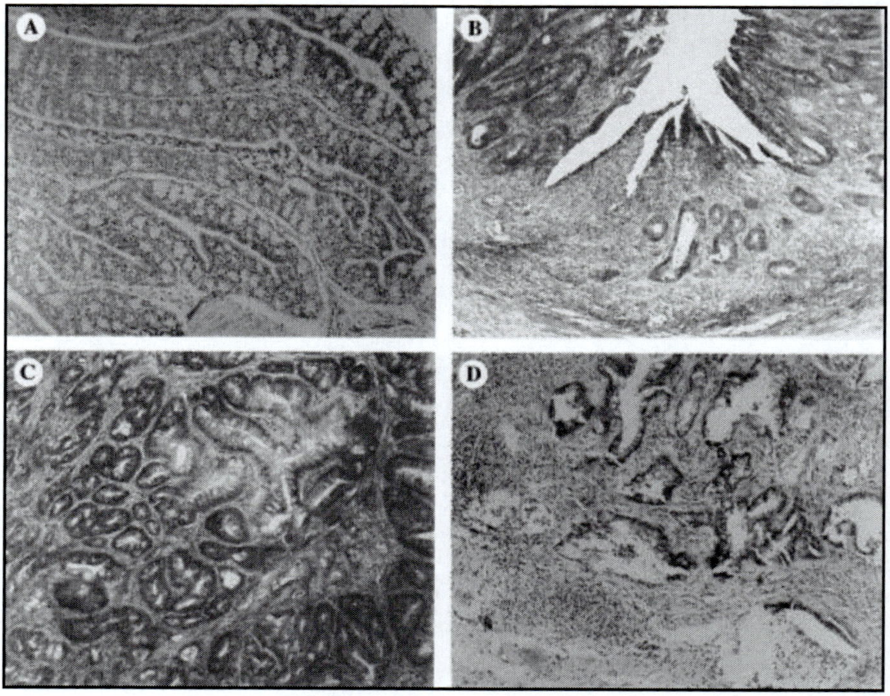

Figure 23. Histological sections from IL-10 knockout mice. Representative gastrointestinal histological sections from IL-10 knockout mice are illustrated. (A) Normal mucosa from an IL-10 knockout mouse consuming the probiotic UCC118. (B) Severe gastrointestinal inflammation from an IL-10 knockout mouse in the placebo group. (C) High grade dysplasia of the colon. (D) Early invasive adenocarcinoma of the colon [70].

Lactobacillus salivarius subsp. *salivarius* UCC118 was detected in faeces from all mice in the probiotic fed group, but not the control group. Faecal coliform and enterococci levels were significantly reduced in probiotic fed animals compared to the controls ($p < 0.05$). At sacrifice, a significant reduction in *Clostridium perfringens* numbers was observed in the test mice ($p < 0.05$). There were no fatalities in the test group compared to two deaths from fulminant colitis in the control group. Only one test mouse developed colonic adenocarcinoma compared to five in the control group. Test animal mucosal inflammation consistently scored lower than that of the control mice. Modification of enteric flora in IL-10 knockout mice by probiotic lactobacilli was associated with reduced prevalence of colon cancer and mucosal inflammatory activity.

Thompson et al., (2001) reported the case of a 2-month-old infant who was born from a 27-year-old mother whose pregnancy was characterized by a lack of prenatal care and cocaine and alchool abuse up to several days before delivery [71]. The immediate postnatal course was complicated by hyaline membrane disease which required intubation until day 11, necrotizing enetrocolitis with jejunal perforation which required bowel resection and a colostomy, and aspiration pneumonia. Blood cultures revealed the presence of *Lactobacillus acidophilus* with a MIC to pennicillin of 0.5 ug/mL and ampicillin of 2 ug/mL so vancomycin and cefotaxime were discontinued and penicillin G and gentamicin were administred intravenously. The patient had no complications and was discharged at 4-1/2 months of age.

The effect on atopic disease of *Lactobacillus* GG was assessed by Kalliomäki et al., (2001) [72]. In a double-blind, randomised placebo-controlled trial *Lactobacillus rhamnosus* GG was given prenatally to mothers who had at least one first-degree relative (or partner) with atopic eczema, allergic rhinitis, or asthma, and postnatally for 6 months to their infants. Chronic recurring atopic eczema, which is the main sign of atopic disease in the first years of life, was the primary endpoint (figure 24 A). Atopic eczema was diagnosed in 46 of 132 (35%) children aged 2 years. Asthma was diagnosed in 6 of these children and allergic rhinitis in one. The frequency of atopic eczema in the probiotic group was half that of the placebo group (15/64 [23%] vs 31/68 [46%]; relative risk 0.51). The number needed to treat was 4.5. *Lactobacillus* GG was effective in prevention of early atopic disease in children at high risk (fig 24 B). Thus, gut microflora could be a hitherto unexplored source of natural immunomodulators and probiotics, for prevention of atopic disease.

Avlami et al., (2001) reported a case of a 65-year-old man that was admitted with a low-grade fever, fatigue, arthralgias, and night sweats of one month's duration [73].

	Placebo (n=82)	***Lactobacillus* GG (n=77)**
Family history		
Maternal atopic disease	63 (77%)	55 (71%)
Atopic eczema in family	35 (43%)	31 (40%)
Older sibling	30 (37%)	26 (34%)
Parental smoking	17 (21%)	9 (12%)
Furry pet at home	9 (11%)	16 (21%)
Birth characteristics		
Weeks of gestation	39 (1·4)	39 (1·3)
Boys	43 (52%)	49 (64%)
Cord blood IgE*	19/63 (30%)	21/53 (40%)
Weight (g)	3612 (466)	3631 (483)
Head circumference (cm)	35 (1·4)	35 (1·3)

Data are mean (SD) or proportion.
*Number of neonates with detectable total IgE (detection limit 0·5 kU/L) in cord blood.

A

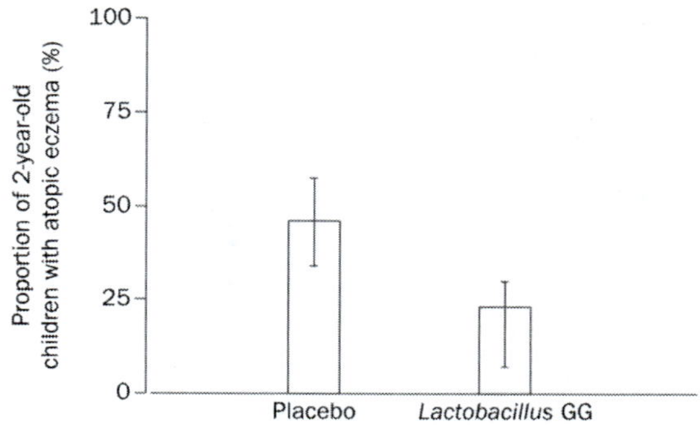

B

Figure 24. (A) Total IgE antibodies, antigen-specific IgE antibodies, and skin-prick test reaction; (B) Treatment effect of *Lactobacillus* GG on atopic disease [72].

The API 50 CHL medium (Biomerieux, La Balmer des Grottes, France) revealed the presence of *Lactobacillus rhamnosus* which was resistant to vancomycin, cefazolin, and imipenem, tolerant to penicillin, rifampin, and erythromycin, relatively resistant to chloramphenicol and gentamicin, and sensitive to clindamycin, ciprofloxacin, cefotaxime, and ceftriaxone. Intravenous ceftriaxone, clindamycin, and ciprofloxacin were used and the fever resolved within 72 hours.

Barton et al., (2001) reported the case of an infant born with gastroschisis who developed *Pediococcus bacteremia* and meningitis 3 months after surgery [74]. Also in this case, it was speculated that such condition was due to the use of probiotics.

Von Der Weid et al., (2001) investigated whether certain strains of LAB could antagonize specific T-helper functions *in vitro* and thus have the potential to prevent inflammatory intestinal immunopathologies [75]. All strains tested induced various levels of both IL-12 and IL-10 in murine splenocytes. In particular, *Lactobacillus paracasei* (strain NCC2461) induced the highest levels of these cytokines. Since IL-12 and IL-10 have the potential to induce and suppress T_h1 functions, respectively, they addressed the impact of this bacterium on the outcome of $CD4^+$ T-cell differentiation. For this purpose, bacteria were added to mixed lymphocyte cultures where $CD4^+$ T-cells from naive BALB/c mice were stimulated weekly in the presence of irradiated allogeneic splenocytes. In these cultures, *Lactobacillus paracasei* NCC2461 strongly inhibited the proliferative activity of $CD4^+$ T cells in a dose-dependent fashion. This was accompanied by a marked decrease of both T_H1 and T_H2 effector cytokines, including INF-γ, IL-4, and IL-5. In contrast, IL-10 was maintained and TGF-β was markedly induced in a dose-dependent manner. The bacteria were not cytotoxic, because cell viability was not affected after two rounds of stimulation. Thus, unidentified bacterial components from *Lactobacillus paracasei* NCC2461 induced the development of a population of $CD4^+$ T cells with low proliferative capacity that produced TGF-β and IL-10.

Gill et al., (2001) conducted a pre–post intervention trial to determine the impact of dietary supplementation with probiotic LAB on peripheral blood NK cell activity in healthy elderly subjects[(76)]. Twenty-seven volunteers consumed low-fat/low-lactose milk supplemented with known immunostimulatory LAB strains (*Lactobacillus rhamnosus* HN001 or *Bifidobacterium lactis* HN019) for a period of 3 weeks. A dietary run-in of milk alone was shown to have no significant effect on NK cells. In contrast, the proportion of CD56-positive lymphocytes in peripheral circulation was higher following consumption of either LAB strain, and *ex vivo* PBMC tumoricidal activity against K562 cells was also increased. Supplementation with HN001 or HN019 increased tumoricidal activity by an average of 101 and 62%, respectively; these increases were significantly correlated with age, with subjects older than 70 years experiencing significantly greater improvements than those under 70 years. These results demonstrated that dietary consumption of probiotic LAB in a milk-based diet could offer benefit to elderly consumers to combat some of the deleterious effects of immunosenescence on cellular immunity.

Liu et al., (2001) examined the effects of *Lactobacillus plantarum* 299v administration on external radiation injury in colon anastomotic healing at different time points [77]. Sprague-Dawley rats were treated with *Lactobacillus plantarum* 299v or saline as control and received external radiation of the lower abdomen (10 Gy/day) on day 3 and 7 of the experiment. After 4 days, a colonic resection with anastomosis was performed. Animals were sacrificed on 4th, 7th and 11th day postoperatively. Body weight, white blood cell (WBC) count, mucosal myeloperoxidase (MPO) activity, hydroxyproline, nucleotide, DNA and RNA content, colonic bacterial microflora, bacterial translocation and histology were evaluated. On the 4th postoperative day body weight, WBC and MPO decreased significantly after radiation. On the 7th postoperative day MPO decreased after radiation. In the two irradiated groups it decreased significantly in the *Lactobacillus plantarum* group compared to the radiated group without treatment. Collagen concentration on the 7th postoperative day was significantly higher in *Lactobacillus plantarum* group without radiation compared to the group with radiation without *Lactobacillus plantarum*. On the 11th postoperative day MPO was significantly higher in irradiated rats without treatment compared to *Lactobacillus plantarum* treatment. The collagen concentration increased significantly in the irradiated *Lactobacillus plantarum* group compared to the other two groups. The collagen content decreased and MPO activity increased significantly of the colonic anastomosis in irradiated rats without treatment compared to those treated with *Lactobacillus plantarum*. It therefore seems that administration of *Lactobacillus plantarum* 299v reduces the intestinal injury and inflammation following external radiation and improves the colonic anastomotic healing.

Sheih et al., (2001) investigated the effects of dietary consumption of *Lactobacillus rhamnosus* HN001 on non-specific polymorphonuclear (PMN) cell phagocytosis and natural killer (NK) cell tumor killing activity [78]. The study was conducted as a three-stage, pre/post trial using healthy middle-aged and elderly volunteers, and its main objective was to determine whether dietary consumption of *Lactobacillus rhamnosus* HN001 could enhance natural cellular immunity. The delivery of HN001 in low-fat milk as a biologically-relevant delivery vehicle was used. The milk was either unaltered or lactose-hydrolyzed; the latter represented an appropriate delivery vehicle for consumers who may develop lactose intolerance. Moreover, lactose-hydrolyzed milk, which is rich in galacto-oligosaccharides, could offer a beneficial prebiotic substrate to optimize any potential immune-enhancing effects of a milk-borne probiotic. The sub-objective of the study was to determine if there was a difference in the immune status of subjects if they consumed HN001 in normal low-fat milk or lactose-hydrolyzed low-fat milk. Immunological responses were unaffected by the run-in diet of LFM alone. In contrast, the relative proportion of PMN cells showing phagocytic activity increased by 19% and 15%, respectively, following consumption of HN001 in either LFM or LFM-LH; the relative level of NK cell tumor killing activity increased by 71% and 147%. In most cases these levels declined following cessation, but remained above baseline. Dietary consumption of *Lactobacillus rhamnosus* HN001, in a base of low-fat milk or lactose-hydrolyzed low-fat milk, appeared to enhance systemic cellular immune responses and could be useful as a dietary supplement to boost natural immunity.

Ulisse et al., (2001) evaluated the tissue levels of pro- and anti-inflammatory cytokines, nitric oxide synthase, and matrix metalloproteinases in control and inflamed pouches before and after antibiotic and probiotic treatment of patients with acute pouchitis [79]. Pouch biopsy samples were obtained from seven patients with pouchitis before and after antibiotic and probiotic treatment. Tissue samples from five patients with normal pouches were used as controls. Cytokines were determined by ELISA, matrix metalloproteinase activity was evaluated by zymograms, and nitric oxide synthase activity was determined by measuring arginine to citrulline conversion. Tissue levels of TNF-α increased ($p < 0.01$) in pouchitis relative to uninflammed pouches and reduced after antibiotic and probiotic treatment. Also, INF-γ and IL-1α augmented in pouchitis, but their increase did not reach statistical significance. The latter, however, were lower ($p < 0.05$) after treatment with the antibiotics and probiotics. Tissue levels of IL-4 and IL-10 were unchanged in inflamed pouches and unaffected by antibiotic treatment. However, IL-10 increased ($p < 0.05$) after probiotic treatment. Moreover, inflamed pouches had higher levels of inducible nitric oxide synthase and gelatinase activities, which decreased after treatment. The ability of antibiotic and probiotic treatments to increase tissue levels of IL-10, at a higher level than those observed in control pouches, and to decrease, to levels present in control pouches, proinflammatory cytokine, inducible nitric oxide synthase, and matrix metalloproteinase activity suggest a mechanism of action to explain the efficacy of this therapeutic regime in pouchitis.

Madsen et al., (2001) determined the efficacy of VSL#3 as a primary therapy in the treatment of colitis in the interleukin IL-10 gene-deficient mouse [80].

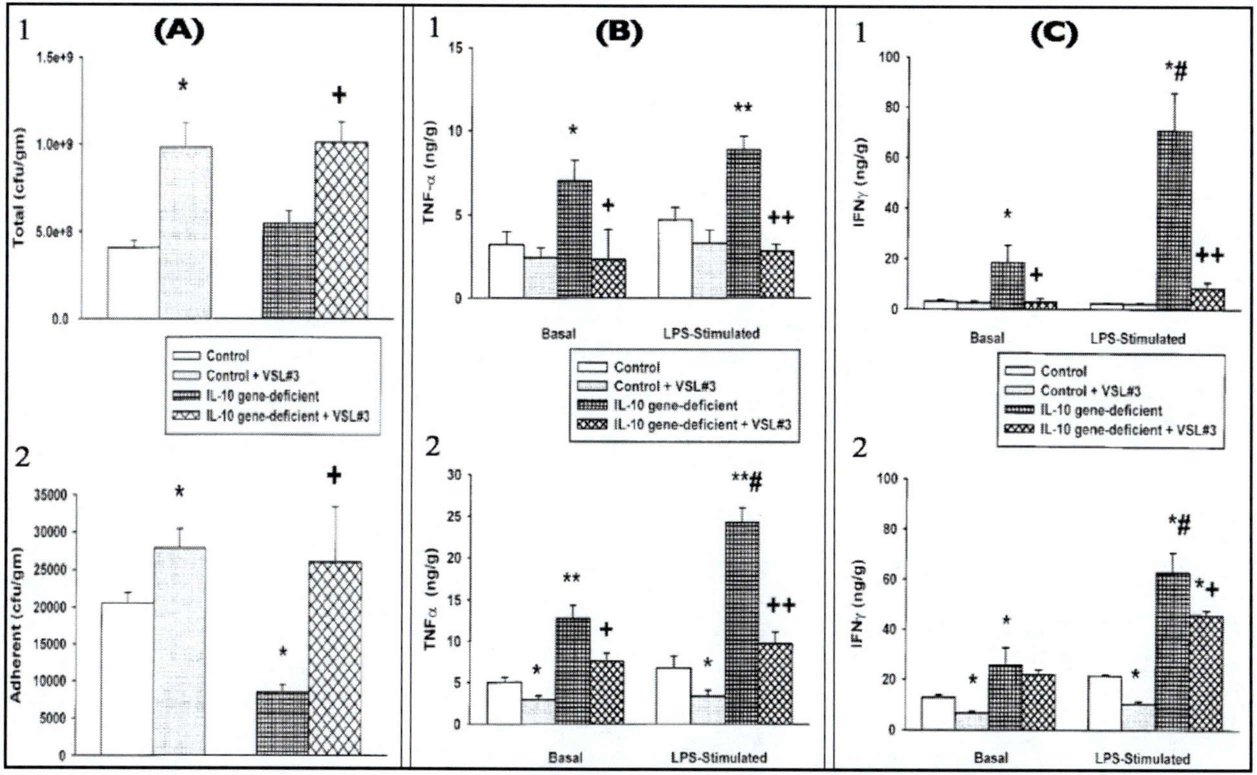

Figure 25. (A) Colonic mucosal (1) total and (2) adherent levels of *Lactobacillus* spp. and *Bifidobacterium* spp. in IL-10 gene-deficient mice (n = 6) and age-matched controls (n = 6). IL-10 gene-deficient mice had significantly reduced levels of adherent *Lactobacillus* spp.

and *Bifidobacterium* spp. compared with controls. Both control mice and IL-10 gene-deficient mice receiving VSL#3 therapy showed increased levels of (1) total and (2) adherent *Lactobacillus* spp. and *Bifidobacterium* spp.; (B) effect of VSL#3 on TNF-α secretion from (1) ileum and (2) colon. Under basal conditions, IL-10 gene-deficient mice (n = 8) secreted significantly higher amounts of TNF-a compared with controls (n = 8) in both the ileum and colon. Colons from IL-10 gene-deficient mice responded to LPS with a significant increase in TNF-α secretion. After 4 weeks of VSL#3 treatment, basal and LPS-stimulated secretion of TNF-α was reduced in the ileum and colon of IL-10 gene-deficient mice. Control mice treated with VSL#3 (n = 8) showed a reduction of basal and LPS-stimulated TNF-α secretion in the colon; (C) effect of VSL#3 on IFN-γ secretion from (1) ileum and (2) colon. Under basal conditions, IL-10 gene-deficient mice (n = 8) secreted significantly higher amounts of IFN-γ compared with controls (n = 8) in both the ileum and colon. The ileum and colon from IL-10 gene-deficient mice responded to LPS with a significant increase in IFN-γ secretion. After 4 weeks of VSL#3 treatment, secretion of basal and LPS-stimulated IFN-γ was reduced in ileum of IL-10 gene-deficient mice. Basal secretion of IFN-γ in the colon was not altered by VSL#3 treatment, but LPS-stimulated secretion was significantly reduced. Control mice treated with VSL#3 (n = 8) showed a reduction of IFN-γ secretion in the colon [80].

Mechanisms of action of VSL#3 were investigated in T84 monolayers. IL-10 gene-deficient and control mice received 2.8 x 10^8 CFU/day of VSL#3 for 4 weeks. Colons were removed and analyzed (figure 25 A) for cytokine production, epithelial barrier function, and inflammation. VSL#3 or conditioned media was applied directly to T84 monolayers. Treatment of IL-10 gene-deficient mice with VSL#3 resulted in normalization of colonic physiologic function and barrier integrity in conjunction with a reduction in mucosal secretion of TNF-α (figure 25 B) and INF-γ (figure 25 C) and an improvement in histologic disease. *In vitro* studies showed that epithelial barrier function and resistance to *Salmonella* invasion could be enhanced by exposure to a proteinaceous soluble factor secreted by the bacteria found in the VSL#3. Oral administration of VSL#3 was effective as primary therapy in IL-10 gene-deficient mice, and had a direct effect on epithelial barrier function (figure 26).

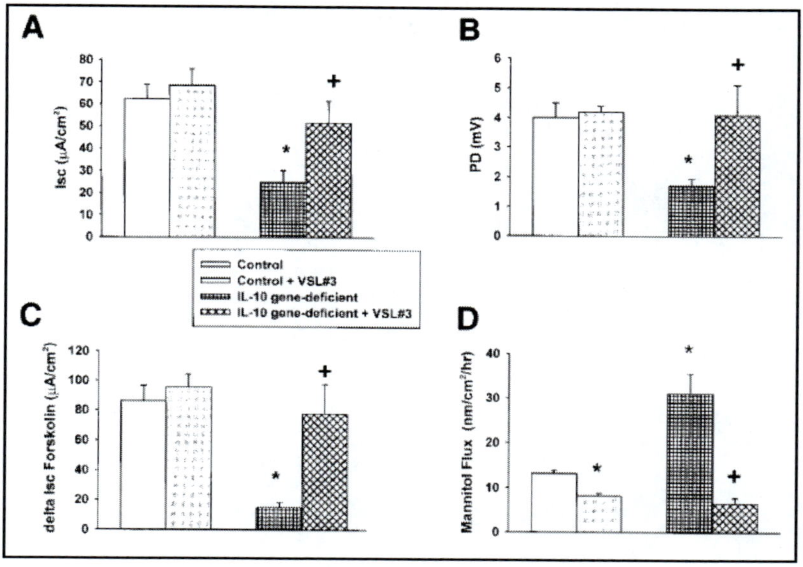

Figure 26. Effects of VSL#3 treatment on colonic electrical parameters and barrier function in control (n = 10) and IL-10–deficient mice (n = 14). IL-10–deficient mice had significantly reduced (A) Isc, (B) PD (potential difference), and (C) D Isc (shortcircuit current) response to forskolin, coupled with an increase in unidirectional (D) mannitol fluxes. Four weeks of VSL#3 therapy normalized all parameters in IL-10 gene-deficient mice. Control mice treated with VSL#3 (n = 6) showed a decrease in mannitol fluxes (D) [80].

Reid and Burton (2002) reviewed the possible immune modulatory mechanisms exerted by *Lactobacillus rhamnosus* GR-1 and B-54 or *Lactobacillus fermentum* RC-14 strains [81]. The instillation of such strains into the vagina had been shown to reduce the risk of urinary tract infections, and improve the maintenance of a normal flora. Ingestion of these strains into the gut had also been shown to modify the vaginal flora to a more healthy state. In addition, these strains inhibited the growth of intestinal, as well as urogenital pathogens, colonized the gut and protected against infections.

Christensen et al., (2002) tested the hypothesis that six different lactobacilli strains (*Lactobacillus reuteri* DSM12246, *Lactobacillus plantarum* Lb1, *Lactobacillus fermentum* Lb20, *Lactobacillus casei* subsp. *alactus* CHCC3137, *Lactobacillus plantarum* 299v, and *Lactobacillus johnsonii* La1) could differentially activate dendritic cells (DC)(82). A concentration of 10 μg/mL of *Lactobacillus casei* induced a high level of IL-12, IL-6, and TNF-α, but almost no IL-10, whereas 100μg/mL induced the same level of IL-12 but a much greater level of IL-10; indicating that lactobacilli induced cytokines production in a dose-dependent fashion at both individual and pattern level. IL-12 and TNF-α were more induced by *Lactobacillus casei* than *Lactobacillus reuteri* as well as for IL-6 and IL-10, although less pronunced. DC cultures, exposed to mixtures of varying concentrations of *Lactobacillus reuteri* and *Lactobacillus casei*, showed an inhibition of *Lactobacillus reuteri* towards IL-12, IL-6, and TNF-α induced by *Lactobacillus casei.* The inhibition was dose-dependent, with maximum inhibition corresponding to a lower limit of cytokine production at the level induced by *Lactobacillus reuteri* alone, indicating a complete abrogation of the *Lactobacillus casei* effect. Even if *Lactobacillus casei* induction of IL-10 was unaffected by *Lactobacillus reuteri*, their combination or LPS exhibited an additive effect on IL-10 induction. Moreover, LPS induction of IL-12, IL-6, TNF-α, or IL-10 was refractory to co-incubation of *Lactobacillus reuteri* at any concentration. Moreover *Lactobacillus casei* showed a stronger capacity to up-regulate MHC class II than the other species of lactobacilli, whereas *Lactobacillus* reuteri showed the weakest capacity. Even at a concentration of 10 μg/mL, *Lactobacillus reuteri* failed to induce MHC class II expression to the same extent as *Lactobacillus casei* at concentrations as low as 1 μg/mL when using the percentage of MHC class II high DC as well as the total percentage of MHC class II low and MHC class II high DC. Similar but more marked trends were found when analyzing for B7-2 up-regulation. While *Lactobacillus reuteri* at 1 g/mL failed to up-regulate B7-2 expression, *Lactobacillus casei* at the same concentration strongly induced B7-2 expression. The effect of

Lactobacillus casei at 1 g/mL on B7-2 expression was even greater than that induced by 10 µg/mL *Lactobacillus reuteri.*

Salminen et al., (2002) evaluated the possible effects of the increased probiotic use of *Lactobacillus rhamnosus* GG on the occurrence of bacteremia due to lactobacilli [83]. Lactobacilli were isolated in 0.02% of all blood cultures and 0.2% of all blood cultures with positive results in Helsinki University Central Hospital and in Finland as a whole, and no trends were seen that suggested an increase in *Lactobacillus* bacteremia (figure 27). The average incidence was 0.3 cases/100.000 inhabitants/year in 1995-2000 in Finland. Identification to the species level was done for 66 cases of *Lactobacillus* bacteremia, and 48 isolates were confirmed to be lactobacilli. Twenty-six of these strains were *Lactobacillus rhamnosus*, and 11 isolates were identical to *Lactobacillus rhamnosus* GG. The results indicate that increased probiotic use of *Lactobacillus rhamnosus* GG has not led to an increase in *Lactobacillus* bacteremia.

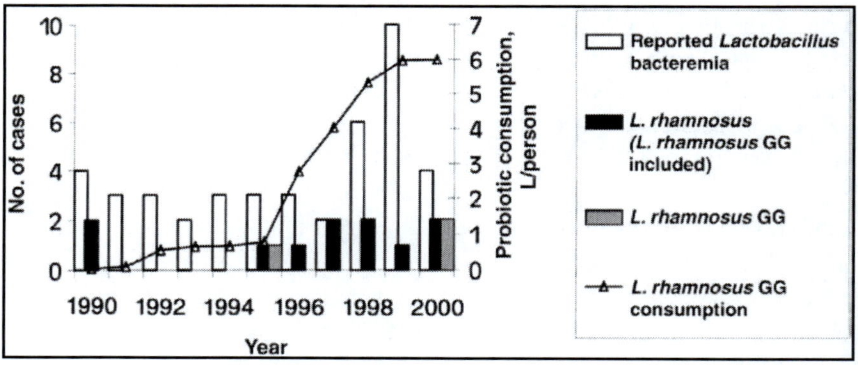

Figure 27. Annual number of cases in which blood culture isolates were reported to be lactobacilli by Helsinki University Central Hospital, 1990-2000, and the annual no. of cases in which the isolate was confirmed to be *Lactobacillus rhamnosus* or *Lactobacillus* rhamnosus GG. The annual consumption in Finland of probiotic *Lactobacillus* GG products in food is also shown [83].

Pochard et al., (2002) studied the effect of LAB on the production of type 2 cytokines, which are known to characterize allergic diseases [84]. They assessed the effect of LAB pre-incubation with PBMCs from patients allergic to house dust mite and with those from healthy donors (figure 28 A). PBMCs were stimulated for 48 hours with the related Dermatophagoides pteronyssinus allergen or with a staphylococcal superantigen. Interestingly, LAB reduced the T_H2 cytokine production from allergic PBMCs specifically re-stimulated with the related allergen (figure 28 B). The inhibition mechanism was dependent on antigen-presenting cells and on the involvement of IL-12 and IFN-γ.

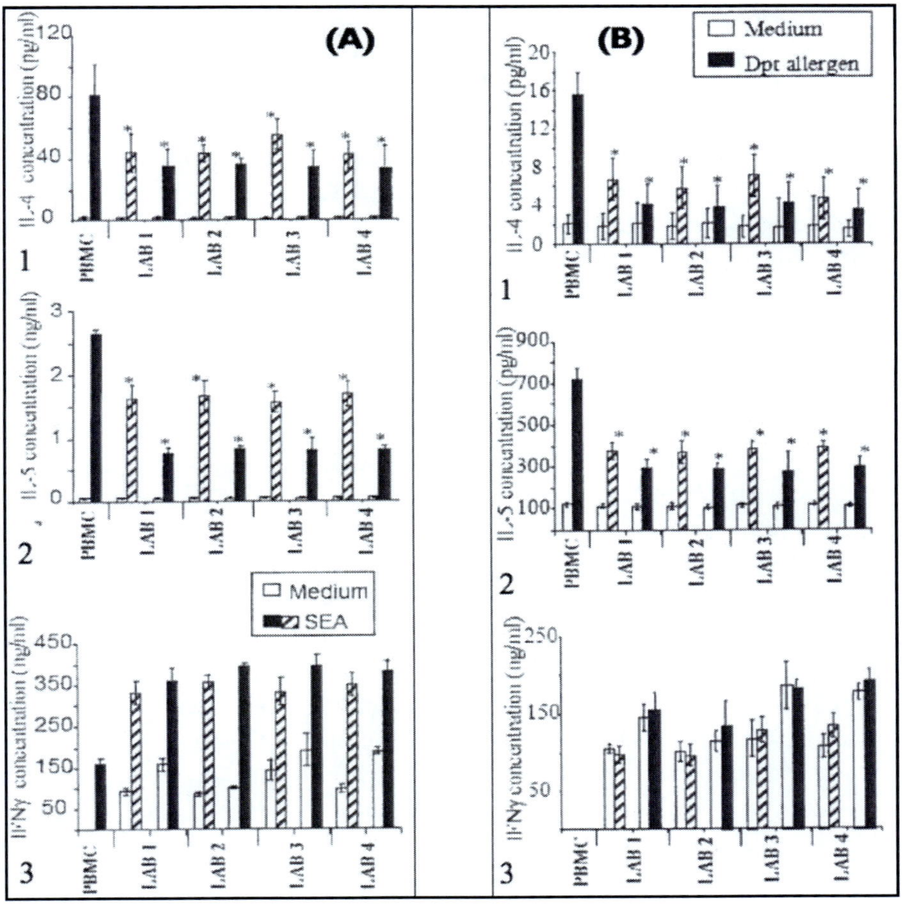

Figure 28. (A) Influence of the concentration of LAB on their immunomodulatory properties. PBMCs from healthy donors (n = 4) were preincubated for 3 h (at 2 x 10^6/mL) with 4 live strains of LAB (LAB1 or *Lactobacillus plantarum* NCIMB8826, LAB2 or *Lactococcus lactis* MG1363, LAB3 or *Lactobacillus casei* ATCC393, and LAB4 or *Lactobacillus rhamnosus* GG) at 2 bacteria/cell ratios (1:1 [striped bars] or 10:1 [solid bars]) before either being stimulated with the SEA superantigen (2 µg/mL) or not; (B) inhibition of allergen-induced T_h2 cytokine profile. PBMCs from patients allergic to *D. pteronyssinus* (Dpt; n = 5) were pre-incubated for 3 hours (at 2 x 10^6/mL) with 4 strains of LAB (LAB1, LAB2, LAB3, and LAB4) at 2 bacteria/cell ratios (1:1 [striped bars] or 10:1 [solid bars]) before related allergen-dependent stimulation (1 IR/mL) or not [84].

McCracken et al., (2002) analyzed the ability of the pro-inflammatory cytokine TNF-α to influence epithelial interleukin (IL)-8 responses to the intestinal bacterium *Lactobacillus plantarum* 299v in the human HT-29 colonic epithelial cell line [85]. In the absence of TNF-α, IL-8 mRNA expression was not detectable by Northern blot analysis in HT-29 cells alone or in HT-29 cells co-cultured with *Lactobacillus plantarum* 299v. However, TNF-α induced IL-8 mRNA expression, and co-culture of TNF-α-treated HT-29 cells with *Lactobacillus plantarum* 299v significantly increased IL-8 mRNA expression above levels induced by TNF-α alone in an adhesion-dependent manner. The increase in IL-8 mRNA expression was not observed in TNF-α-treated HT-29/*Lactobacillus plantarum* 299v co-cultures using heat-killed lactobacilli or when *Lactobacillus plantarum* adhesion was prevented using mannoside or a trans-well membrane. Paradoxically, IL-8 secretion was decreased in TNF-α- treated HT-29 cells with *Lactobacillus plantarum* 299v relative to cells treated with TNF-α alone. TNF-α-mediated responsiveness to *Lactobacillus plantarum* 299v was further investigated by analyzing expression of a co-receptor for bacterial cell wall products CD14. HT-29 cells expressed CD14 mRNA and cell-surface CD14; however, TNF-α did not alter CD14 mRNA or cell-surface expression, and blockade of CD14 with monoclonal antibody MY4 did not alter the IL-8 response to *Lactobacillus plantarum* 299v in TNF-α-treated HT-29 cells. The results indicated that although TNF-α sensitized HT-29 epithelial cells to intestinal lactobacilli, the bacteria exerted a protective effect by downregulating IL-8 secretion.

Cukrowska et al., (2002) analyzed the influence of oral administration of *Escherichia coli* Nissle 1917 on the systemic humoral and cellular immunity in premature infants [86]. Thirty-four premature infants were colonized with *Escherichia coli* Nissle 1917 in a randomized, placebo-controlled blinded clinical trial. Stool samples of infants were analyzed repeatedly for the presence of the administered strain. The proliferative response to bacterial antigens of *Escherichia coli* origin was measured in whole blood of 34 colonized infants and 27 noncolonized controls (figure 29). *Escherichia coli* colonization induced a significant increase in the proliferation of blood cells cultivated with bacterial components of *Escherichia coli* Nissle 1917 and another *Escherichia coli* strain in colonized infants as compared with noncolonized controls. Significantly higher amounts of specific anti-*Escherichia coli* Nissle 1917 antibodies (Ab) of immunoglobulin (Ig)A isotype and nonspecific polyclonal IgM were found in the blood of colonized infants compared to noncolonized placebo controls. The oral application of *Escherichia coli* Nissle 1917 after birth significantly stimulated specific humoral and cellular responses and simultaneously induced nonspecific natural immunity.

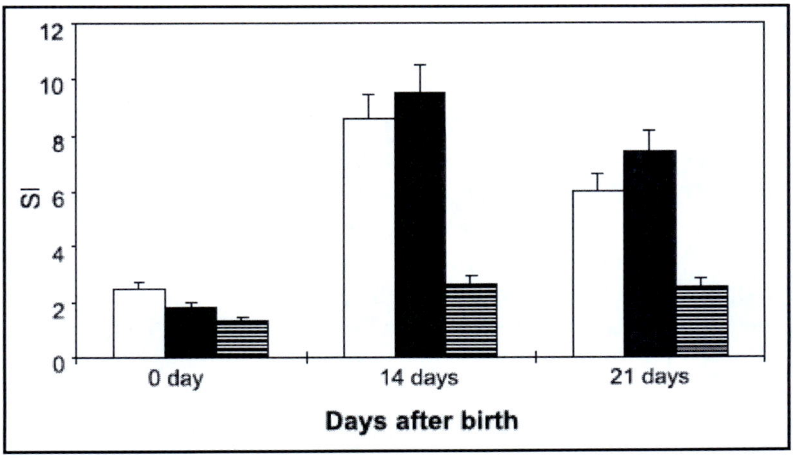

Figure 29. Dynamics of the cellular response in blood of premature infants (n = 14) after the oral application of *Escherichia coli* strain Nissle 1917. The cellular response was measured by a proliferative assay of whole blood before colonization (day 0) in cord blood (CB), 14 and 21 days after colonization in the blood of premature infants. The blood was incubated for 5 days with the following antigens: heatinactivated *Escherichia coli* Nissle 1917 (white bars), sonicate from this *Escherichia coli* strain (black bars) and lipopolysaccharide (LPS) isolated from *Escherichia coli* O86 (hatched bars). The results are expressed as SI, i.e., the mean radioactive counts per minute with the antigen divided by the mean counts without antigen (with medium only). Data (n = 14) are presented as arithmetical mean ± SE [86].

He et al., (2002) cultured a murine macrophage-like cell line (J774.1), in the presence of 27 strains of heat-inactivated Bifidobacteria, in order to characterize the ability of Bifidobacteria to affect the production of macrophage-derived cytokines [87]. *Bifidobacterium adolescentis* and *Bifidobacterium longum*, known as adult-type Bifidobacteria, induced significantly more pro-inflammatory cytokine secretion, IL-12 and TNF-α, by J774.1 cells, than did the infant-type Bifidobacteria, *Bifidobacterium bifidum*, *Bifidobacterium breve*, and *Bifidobacterium infantis* ($p < 0.01$). In contrast, *Bifidobacterium adolescentis* did not stimulate the production of anti-inflammatory IL-10 from J774.1 cells as the other tested bacteria did. The adult-type Bifidobacteria, especially *Bifidobacterium adolescentis*, could be more potent to amplify but less able to down-regulate the inflammatory response.

Perdigón et al., (2002) evaluated the antigenic effect of LABs (*Lactobacillus acidophilus*, *Lactobacillus casei*, *Lactobacillus delbrueckii* ssp. *bulgaricus* and *Streptococcus thermophilus*) on the gut immune system of BALB/c mice [88]. A dose dependent increase of the Bcl2 protein was observed with all LAB assayed (figure 30 A). Moreover the analysis of cytokine-producing cells in the lamina propria of gut showed that TNF-α and INF-γ values, determined in macrophages cultured from Peyer patches, were enhanced for all the LABs assayed. An important increase of interleukins IL-10 and IL-4 was observed mainly in mice fed with *Lactobacillus delbrueckii* ssp. *bulgaricus* or *Lactobacillus casei*, while a significant induction of IL-2 and IL-12 was only observed with *Lactobacillus acidophilus* (figure 30 B). These effects were dose dependent. The role of produced cytokines in the balance T_H1/T_H2 was determined by a systemic antibody response against parenterally injected ovoalbumin (figure 30 C). *Lactobacillus casei*, *Lactobacillus delbrueckii* subsp. *bulgaricus* and *Lactobacillus acidophilus* enhanced the IgG1 response favouring T_H2 balance, while *Lactobacillus acidophilus* also increased the IgG2a response inducing T_H1 balance. *Steptococcus thermophilus* did not influence the balance $T_H1=T_H2$. LAB induced distinct mucosal cytokine profiles showing different adjuvant capacity among them.

Cross et al., in 2002, investigated the ability of a known IFN-γ-promoting probiotic (*Lactobacillus rhamnosus* HN001) to modulate cytokine production in mice expressing an on-going T_H2-type immune response [89]. BALB/c mice were primed to ovalbumin in alum adjuvant to invoke antigen-specific T_H2 cytokine-secreting cell populations. Mice that were fed *Lactobacillus rhamnosus* HN001 during antigen sensitization produced higher levels of lymphocyte-derived IFN-γ, but also IL-4 and IL-5, in comparison to control animals. Although HN001 was additionally shown to induce pro-IFN-γ monokine (IL-12, IL-18) secretion in macrophages *in vitro*, its ability to invoke mixed lymphocyte cytokine production during an on-going T_H2-type immune response *in vivo* suggested that this probiotic was a general immunostimulatory agent, in contrast to the pro-T_H1/anti-T_H2 immunoregulation reported for some strains of IFN-γ-promoting lactobacilli.

Borruel et al., (2002) investigated the effect of bacteria on TNF-α release by intestinal mucosa. Ileal specimens were obtained at surgery from 10 patients with Crohn's disease (ileal stricture) and five disease controls undergoing right hemicolectomy (caecal cancer) [90].

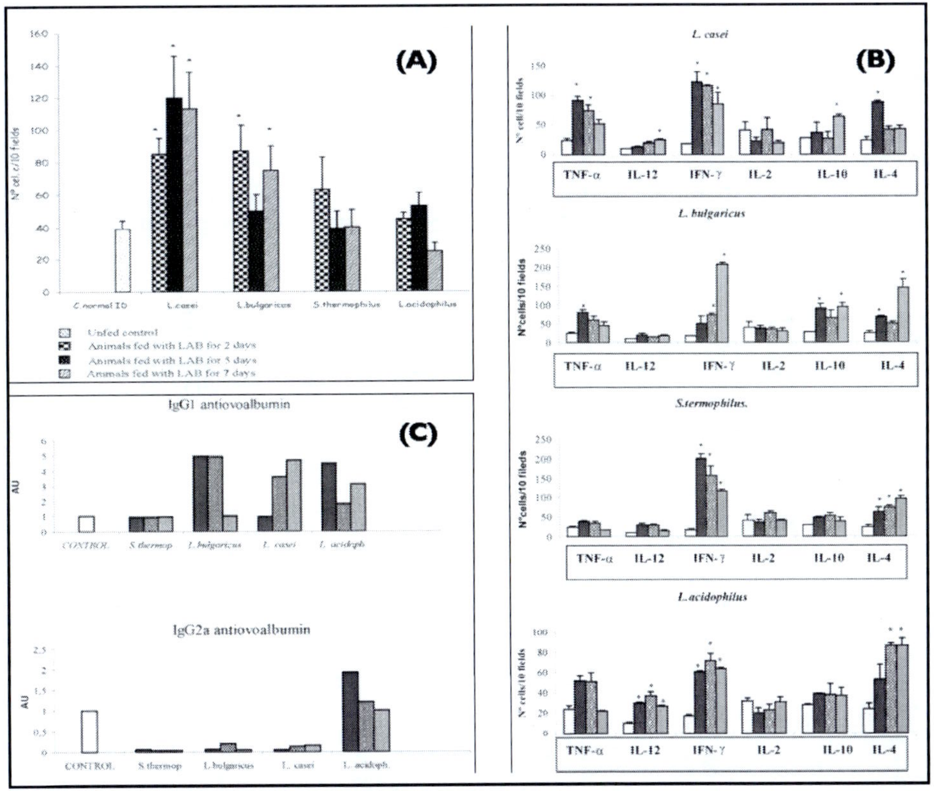

Figure 30. (A) Effect of orally administered LAB on the induction of Bcl2 protein. LAB were administered for 2, 5 or 7 days. At the end of each period feeding, small intestine Bcl2 protein was determined on histological slices by immunofluorescence test. (Continued)

Values are the means of n mice= 3± s.d. $*p < 0.01$; (B) production of cytokines by LAB orally administered LAB were given for 2, 5 or 7 days; (C) influence of LAB administration on the $T_H1 = T_H2$ population IgG1 (T_H2 response) or IgG2a (T_H1 response) antiovoalbumin was determined by ELISA test in the sera of animals intraperitoneally injected with the antigen which had been previously fed with the different LABs [88].

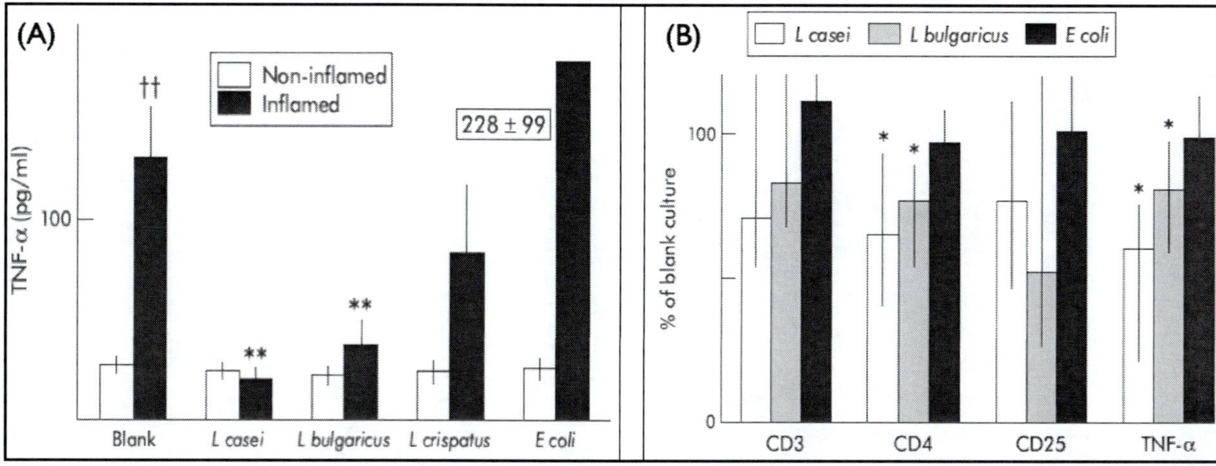

Figure 31. (A) Levels of TNF-α in the incubation medium after 24 h of culture of intestinal mucosa from Crohn's disease patients in macroscopically non-inflamed mucosa or inflamed mucosa. No bacterium was added to blank cultures whereas *Lactobacillus casei*, *Lactobacillus bulgaricus*, *Lactobacillus crispatus*, or *Escherichia coli* were added at 10^6 CFU/mL in the corresponding culture wells. In blank cultures, levels of TNF-α were significantly higher in inflamed than in non-inflamed mucosa. Co-culture of inflamed mucosa with *Lactobacillus casei* or *Lactobacillus bulgaricus* significantly reduced TNF-α levels: (B) phenotype of intraepithelial lymphocytes from inflamed mucosa of Crohn's disease patients normalized as a percentage of the data in blank cultures of mucosa from the same individual (100%). Co-culture with *Lactobacillus casei* or *Lactobacillus bulgaricus* reduced $CD4^+$ cell numbers, as well as expression of TNF-α. Co-culture with *Escherichia coli* had no effect on CD3, CD4, CD25, or TNF-α [90].

Mucosal explants from each specimen were cultured for 24 hours with either non-pathogenic *Escherichia coli*, *Lactobacillus casei* DN-114001, *Lactobacillus bulgaricus* LB10, or *Lactobacillus crispatus* (each study contained blank wells with no bacteria). Tissue and bacterial viability was confirmed by lactate dehydrogenase (LDH) release and culture. Concentrations of TNF-α were measured in supernatants and the phenotype of the intestinal lymphocytes was analysed by flow cytometry (figure 31 A). Co-culture of mucosa with bacteria did not modify LDH release. Release of TNF-α by inflamed Crohn's disease mucosa was significantly reduced by co-culture with *Lactobacillus casei* or *Lactobacillus bulgaricus*; changes induced by *Lactobacillus crispatus* or *Escherichia coli* were not significant. The effect of *Lactobacillus casei* and *Lactobacillus bulgaricus* was not prevented by protease inhibitors. Coculture with *Lactobacillus casei* and *Lactobacillus bulgaricus* reduced the number of CD4 cells as well as TNF-α expression among intraepithelial lymphocytes from Crohn's disease mucosa (figure 31 B). None of the bacteria induced changes in non-inflamed mucosa. Probiotics interact with immunocompetent cells using the mucosal interface and modulate locally the production of pro-inflammatory cytokines.

Schultz et al., (2002) used four different protocols that varied the time-points of oral administration of *Lactobacillus plantarum* 299v relative to colonization with SPF bacteria to determine whether *Lactobacillus plantarum* could prevent and treat colitis induced by SPF bacteria in IL-10-/- mice and evaluated the effect of this probiotic organism on mucosal immune activation [91]. Assessment of colitis included blinded histologic scores, measurements of secreted colonic immunoglobulin isotypes, IL-12 (p40 subunit), and IFN-γ production by anti-CD3-stimulated mesenteric lymph node cells. Treating SPF IL-10-/- mice with *Lactobacillus plantarum* attenuated previously established colonic inflammation as manifested by decreased mucosal IL-12, IFN-γ, and immunoglobulin G2a levels. Colonizing GF animals with *Lactobacillus plantarum* and SPF flora simultaneously had no protective effects. Gnotobiotic IL-10-/- mice monoassociated with *Lactobacillus plantarum* exhibited mild immune system activation but no colitis. Pretreatment of GF mice by colonization with *Lactobacillus plantarum*, then exposure to SPF flora and continued probiotic therapy significantly decreased histologic colitis scores. *Lactobacillus plantarum* was able to attenuate immune-mediated colitis suggesting a potential therapeutic role for this agent in clinical inflammatory bowel diseases.

Borruel et al., (2003) tried to observe whether host-bacteria interactions influence mucosal cytokine production [92]. Macroscopically normal colonic specimens were obtained at surgery from eight patients with neoplasm, and inflamed ileal specimens were obtained from two patients with Crohn's disease. Mucosal explants were cultured for 24 hours with either non-pathogenic *Escherichia coli* ECOR-26, *Lactobacillus casei* DN-114 001, *Lactobacillus casei* DN-114 056, *Lactobacillus casei* ATCC-334, or *Lactobacillus bulgaricus* LB-10. Each study included blank wells with no bacteria. Tissue and bacteria viability were confirmed by LDH release and culture. Concentration of TNF-α, TGF-β1, IL-8, and IL-10 was measured in supernatants. In parallel experiments, neutralizing anti-TNF-α antibody was added to the culture. Co-culture of mucosa with bacteria did not modify LDH release. Co-culture with *Lactobacillus casei* strains significantly reduced TNF-α release, whereas *Escherichia coli* increased it. These effects were observed both in normal and inflamed mucosa. In combination studies, *Lactobacillus casei* DN-114 001 prevented TNF-α stimulation by *Escherichia coli. Lactobacillus casei* DN-114 001 also reduced IL-8 release via a TNF-α-independent pathway. *Lactobacillus casei* DN-114 056 or *Escherichia coli* increased IL-10 release in the presence of neutralizing anti-TNF-α. Nonpathogenic bacteria interact with human intestinal mucosa and can induce changes in cytokine production that are strain specific.

McCarthy et al., (2003) compared the effect of feeding *Lactobacillus salivarius* subspecies *salivarius* 433118 and *Bifidobacterium infantis* 35624 against placebo on enterocolitis (figure 32), the intestinal microflora, and the systemic immunological response to *in vitro* microbial challenge in probiotic fed and control IL-10 KO mice [93]. Three groups of 10 IL-10 KO mice were fed fermented milk products containing *Lactobacillus salivarius* 433118 at 10^9 CFU/mL, *Bifidobacterium infantis* 35624 at 10^8 CFU/mL, and unmodified milk, respectively, for 19 weeks. Faecal samples were taken at regular intervals to confirm gut transit, recovery of fed probiotics, and to assess the impact on the microflora. At sacrifice, the bowels were histologically scored. Cytokine production from Peyers' patches and splenocytes was measured *in vitro* by ELISA. Faecal recovery of probiotics was confirmed in all probiotic fed mice but not in controls. Colonic and caecal inflammatory scores were significantly decreased in both groups of probiotic fed mice ($p < 0.05$). Proinflammatory cytokine production by Peyers' patches and splenocytes was significantly reduced in probiotic fed animals whereas transforming growth factor β (TGF-β) levels were maintained. Thus both *Lactobacillus salivarius* 433118 and *Bifidobacterium infantis* 35624 significantly attenuated colitis in this murine model. Attenuation of colitis was associated with a reduced ability to produce T_H1-type cytokines systemically and mucosally, while levels of TGF-β were maintained.

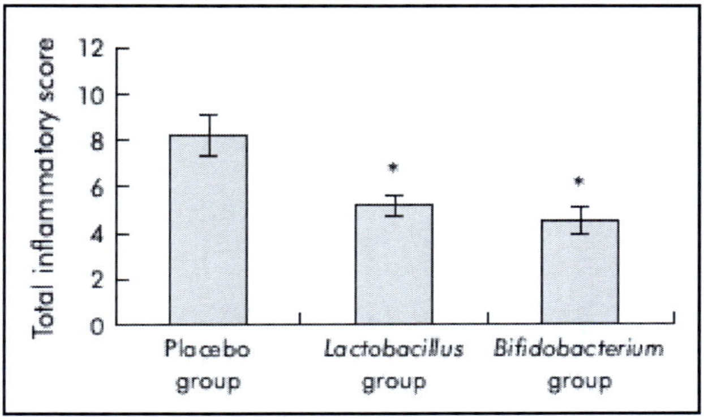

Figure 32. Gastrointestinal inflammatory score, including the ileum, caecum, and colon, in probiotic and control fed groups (*$p < 0.05$; n=10 each group). Results are expressed as mean (SEM) inflammatory score [93].

Riquelme et al., (2003) reported two cases of fungemia caused by *Saccaromyces cerevisiae* occurring in immunosuppressed patients treated orally with Saccaromyces boulardii (a 42-year-old woman positive for *Clostridium difficile* presence and a 41-year-old man negative for *Clostridium difficile* presence) [94]. They both confirmed, by molecular typing, clonality in isolate strains from patients and the *Saccharomyces boulardii* capsule and reported other cases of *Saccharomyces cerevisiae* fungemia which were treated with fluconazole and amphotericin B. They also reported other cases of fungemia in literature (figure 33).

Lammers et al., (2003) incubated peripheral blood mononuclear cells (PBMC) from healthy donors with pure DNA of eight probiotic strains and with total bacterial DNA from human feces collected before and after probiotic ingestion (figure 34) [95]. Cytokine production was analyzed in culture supernatants. Modification of human microflora after probiotic administration was proven by polymerase chain reaction analysis. *Bifidobacterium* genomic DNA induced secretion of the anti-inflammatory IL-10 by PBMC. Total bacterial DNA from feces collected after probiotic administration modulated the immune response by a decrease of IL-1L and an increase of IL-10.

Number of cases	Age (year)	Clinical characteristics and immunological status	*Clostridium difficile* diarrhea	Terapy	Survival
1	51	Vasculitis, Pneumocystis caninii and cytomegalovirus. Immunosuppressive therapy with corticosteroids and cyclophosfamide	Present	Amphotericin B	Yes
1	1	Chronic diarrhea	Absent	Fluconazole	Yes
1	Neonate	Congenital cardiopathy	Absent	Amphotericin B	Yes
1	74	Subarachnoid hematoma and peritonitis	Absent	Fluconazole	No*
1	78	Critically ill patient, chronic pulmonary disease, and nosocomial pneumonia	Absent	Fluconazole	Yes
1	49	Aspirative pneumonia under steroid treatment	Absent	Fluconazole	Yes
1	8 months	Acute leukemia with neutropenia. Recent chemotherapy	Absent	Amphotericin B and Fluconazole	Yes
1	14	Severe burn	Absent	Amphotericin B	Yes
1	33	Ulcerative colitis under steroid treatment	Absent	Amphotericin B and Fluconazole	Yes

Figure 33. Summary of published reports of *Saccharomyces cerevisiae* fungemia after oral therapy with *Saccharomyces boulardii* (Modified from Riquelme et al., 2003) [94].

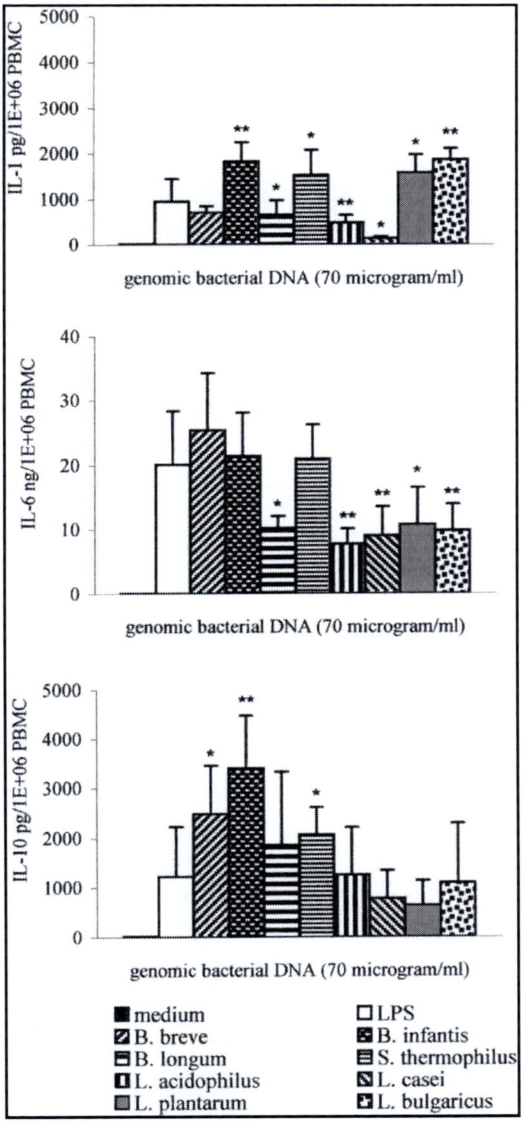

Figure 34. PBMC from four healthy donors were cultured with 70 μg/mL of each genomic DNA extracted from VSL#3 bacteria. IL-1L, IL-6 and IL-10 production was measured in supernatants. The cytokine response to bacterial DNAs was compared to that obtained with LPS (20 ng/mL): $*p < 0.05$, $** p < 0.01$ [96].

Benyacoub et al., (2003) studied the capacity of a probiotic LAB to stimulate immune functions in young dogs [96]. Puppies were allotted to two groups receiving either a control diet or a diet supplemented with 5 x 10^8 CFU/d of probiotic *Enterococcus faecium* (SF68) from weaning to one year of age. Fecal and blood samples were collected from the dogs at different time points for the measurement of fecal (Ig)A, circulating IgG and IgA, and the proportions of lymphoid cell subsets. Fecal IgA and canine distemper virus (CDV) vaccine-specific circulating IgG and IgA were higher in the group receiving the probiotic than in controls. There were no differences in the percentages of CD4 and CD8 T cells between the groups (figure 35), but the proportion of mature B cells [CD21/major histocompatibility complex (MHC) class II] was greater in those fed the probiotic. Data showed for the first time that a dietary probiotic LAB enhance specific immune functions in young dogs, thus offering new opportunities for the utilization of probiotics in canine nutrition.

Lymphocyte subsets	Treatment	Week 18	Week 31	Week 44
$CD4^+$ T cells	Control	32.3 ± 2.5	32.7 ± 3.7	37.0 ± 5.6
	E. faecium	33.0 ± 4.3	35.2 ± 3.0	39.0 ± 4.0
$CD8^+$ T cells	Control	18.3 ± 3.2	14.5 ± 2.5	18.3 ± 1.4
	E. faecium	18.0 ± 2.8	15.6 ± 0.6	16.8 ± 1.0
$CD4^+/CD8^+$ T cells	Control	1.7 ± 0.4	2.2 ± 0.8	2.0 ± 0.2
	E. faecium	1.8 ± 0.6	2.2 ± 0.4	2.3 ± 0.3
$CD21^+/MHCII^+$ B cells	Control	18.1 ± 4.2	10.2 ± 1.2	11.5 ± 1.5
	E. faecium	23.3 ± 4.7	17.6 ± 1.8*	16.0 ± 1.6*

Figure 35. Percentage of different lymphoid cell subsets in growing puppies fed *Enterococcus faecium* [96].

Braat et al., (2004) determined whether *Lactobacillus rhamnosus* could modulate T cell responses *in vitro* and *in vivo* [97]. *In vitro*, human monocyte-derived dendritic cells (DCs) matured in the presence of *Lactobacillus rhamnosus* were used to instruct naive $CD4^+$ T cells; subsequently, the T cell response was assessed with the use of CD3/CD28 and IL-2. Cytokine production by *ex vivo*–stimulated naive cells and memory T cells was measured before and after oral supplementation with *Lactobacillus rhamnosus* in six healthy volunteers and 6 patients with Crohn disease. A decreased T cell proliferation and cytokine production, especially of IL-2, IL-4, and IL-10, was observed in CD3/CD28- stimulated T cells derived from *Lactobacillus rhamnosus* – matured DCs (figure 36). This T cell hyporesponsiveness was associated with enhanced DC–T cell interaction and normal responsiveness of T cells for IL-2. *In vivo* oral supplementation of *Lactobacillus rhamnosus* for 2 weeks induced a similar T cell hyporesponsiveness, including impaired *ex vivo* T helper subsets 1 and 2 responses without up-regulation of immunoregulatory cytokines in cohorts of both healthy volunteers and patients with Crohn disease. They proposed that *Lactobacillus rhamnosus* modulated DC function to induce a novel form of T cell hyporesponsiveness; this mechanism could be an explanation for the observed beneficial effects of probiotic treatment in clinical disease.

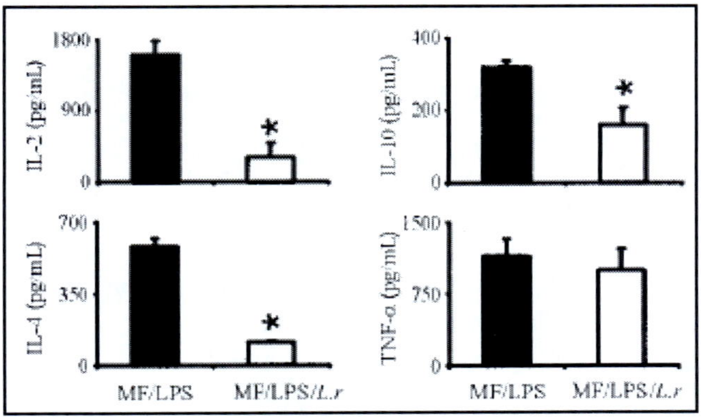

Figure 36. Mean (± SEM) interleukin (IL) 2, 4, and 10 and TNF-α concentrations after the 48-h CD3/CD28 stimulation of dendritic cell (DC)–primed T cells. Immature DCs were stimulated with TNF-α and IL-1β [maturation factors (MF)] and lipopolysaccharide (LPS) (MF/LPS) or MF, LPS, and Lactobacillus rhamnosus (MF/LPS/L.r) for 48 h. Subsequently, mature DCs were washed twice, counted, and seeded (5000 cells/well) with naive T cells (25 000/well) until full proliferation was achieved. T cells were then harvested, seeded (100 000 cells/well) in triplicate, and stimulated for 48 h with CD3 and CD28; supernatant fractions were used for the cytokine measurements. *Significantly different from MF/LPS, $p = 0.05$ (heteroscedastic Student's t test) [97].

Jijon et al., (2004) examined whether bacterial DNA (*Salmonella dublin* strain Lane ATCC 15480, *Salmonella typhimurium* SL1344, *Enterococcus faecalis* ATCC 49149, *Lactobacillus acidophilus* MB 443, *Lactobacillus delbrueckii* subsp. *bulgaricus* MB 453, *Lactobacillus casei* MB 451, *Lactobacillus plantarum* MB 452, *Bifidobacterium longum* Y10, *Bifidobacterium infantis* Y1, *Bifidobacterium breve* Y8, and *Streptococcus salivarius* subsp. *thermophilus* MB 455), could serve as the molecular basis for bacterial recognition(98). HT-29 monolayers were treated with various bacterial DNA and IL-8 secretion measured by ELISA (figure 41), NFκB activation by electrophoretic mobility shift assay and reporter assays, and IκB levels by Western blotting. Cytokine secretion in response to bacterial DNA was measured in murine colonic segments and splenocytes. IL-10–deficient mice were fed DNA from VSL#3 probiotic compound daily for 2 weeks. Colons were removed and analyzed for cytokine production and inflammation. HT-29 cells responded with IL-8 secretion to bacterial DNA in a differential manner. In the presence of pro-inflammatory stimuli, VSL#3 DNA inhibited IL-8 secretion, reduced p38 mitogen-activated protein kinase activation, delayed NFκB activation, stabilized levels of IκB, and inhibited proteasome function. VSL#3 DNA inhibited colonic IFN-γ secretion in mouse colons and also attenuated a Bacteroides vulgatus-induced IFN-γ release from murine splenocytes. In mice, VSL#3 DNA attenuated a systemic release of TNF-α in response to *Escherichia coli* DNA injection. Treatment of IL-10–deficient mice with oral VSL#3 DNA (figure 37) resulted in a reduction in mucosal secretion of TNF-α and IFN-γ and an improvement in histologic disease. DNA from probiotic bacteria could limit epithelial pro-inflammatory responses *in vivo* and *in vitro*. Systemic and oral administration of VSL#3 DNA ameliorated inflammatory responses.

Kunz et al., (2004) reported two cases of *Lactobacillus* GG sepsis occurring during the therapeutic use of this organism [99]. Particularly when associated with intestinal friability, patients could be vulnerable to bacteremia or sepsis.

Panigrahi et al., (2004) examined the effect of supplementing a suggested probiotic bacteria *Lactobacillus rhamnosus* JCM 1136 in feed on immune response and gut flora composition of rainbow trout *Oncorhynchus mykiss* [100]. The probiotic bacteria were incorporated into a commercial feed to constitute two experimental diets containing either 10^9 or 10^{11} CFU of live bacteria/g of feed while a third diet without the bacterial supplement served as the control diet. The diets were offered to rainbow trout (75 g average weight) in triplicate tanks for 30 days. Fish were sampled at 10, 20 and 30 days after commencement of the feeding trial to determine the proportion of the given probiont in the gut microflora composition and the nonspecific humoral and cellular immune responses on the 30th day. The relative proportion of the probiont increased with the feeding duration in the intestine, but not in the stomach. The proportion of *Lactobacillus rhamnosus* in the stomach corresponded to the intake levels while no such relation existed in the intestine. The serum lysozyme and complement activities were significantly greater in fish fed the higher level of probiont compared with the control fish.

The phagocytic activity of head kidney leucocytes also showed similar tendencies. These observations indicated the potential immuno-regulatory role of probiotic organisms in rainbow trout.

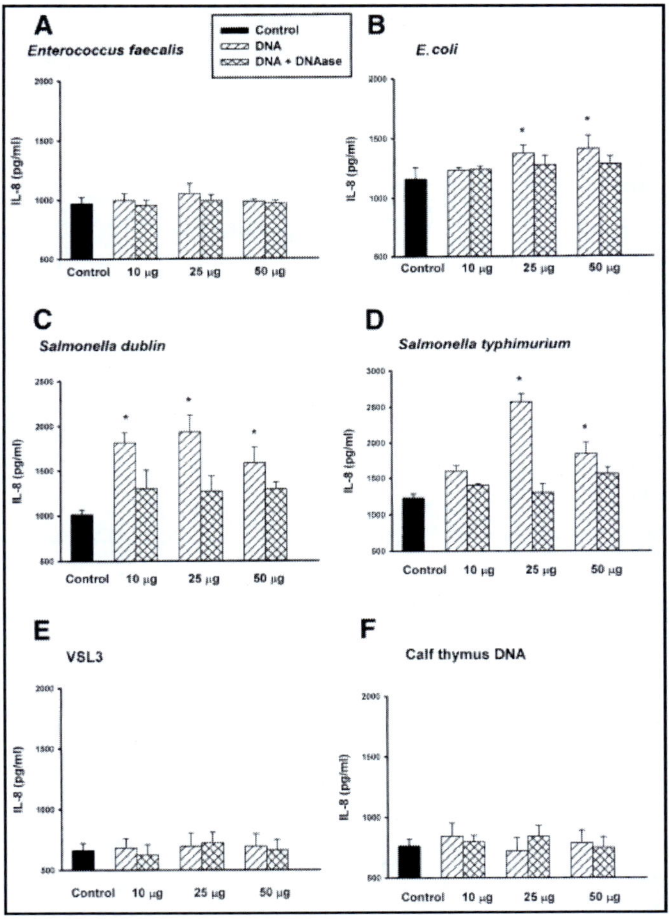

Figure 37. Effect of bacterial and calf thymus DNA on IL-8 secretion from HT-29 monolayers [98].

Schultz et al., (2004) evaluated murine models of acute and chronic intestinal inflammation to study the antiinflammatory effect of EcN (*Escherichia coli* strain Nissle 1917) *in vivo* [101]. Acute colitis was induced in mice with 2% dextran-sodium sulfate (DSS) in drinking water. EcN was administered from day - 2 to day + 7. Chronic colitis was induced by transfer of $CD4^+$ $CD62L^+$ T lymphocytes from BALB/c mice in SCID mice. EcN was administered three times/week from week 1 to week 8 after cell transfer. Mesenteric lymph node (MLN) cytokine secretion (of IFN-γ, IL-5, IL-6, and IL-10) was measured by enzyme-linked immunosorbent assay (ELISA). Histologic sections of the colon were analyzed by using a score system ranging from 0 to 4. Intestinal contents and homogenized MLN were cultured, and the number of *Escherichia coli*-like colonies was determined. EcN was identified by repetitive extragenic palindromic (REP) PCR. EcN administration to DSS-treated mice reduced the secretion of proinflammatory cytokines (IFN-γ, 32.477 ± 6.377 versus 9.734 ± 1.717 [$p = 0.004$]; IL-6. 231 ± 35 versus 121 ± 17 [$p = 0.02$]) but had no effect on the mucosal inflammation. In the chronic experimental colitis of the transfer model, EcN ameliorated the intestinal inflammation (histology score, 2.7 ± 0.2 versus 1.9 ± 0.3 [$p = 0.02$]) and reduced the secretion of proinflammatory cytokines. Translocation of EcN and resident *Escherichia coli* into MLN was observed in the chronic colitis model but not in healthy controls. Administration of EcN ameliorated acute and chronic experimental colitis by modifying proinflammatory cytokine secretion but had no influence on the acute DSS-induced colitis.

Wehkamp et al., (2004) compared probiotic bacterial strains such as *Escherichia coli* Nissle 1917 with nonprobiotic, pathogenic and nonpathogenic bacteria with respect to innate defense mechanisms in the intestinal mucosal cell [102]. *Escherichia coli* Nissle 1917 and a variety of other probiotic bacteria, including lactobacilli - in contrast to more than 40 different *Escherichia coli* strains tested-strongly induced the expression of the antimicrobial peptide human beta-defensin-2 (hBD-2) in CaCO-2 intestinal epithelial cells in a time- and dose-dependent manner. Induction of hBD-2 through *Escherichia coli* Nissle 1917 was further confirmed by activation of the hBD-2 promoter and detection of the hBD-2 peptide in the culture supernatants of *Escherichia coli* Nissle 1917-treated Caco-2 cells. Luciferase gene reporter analyses and site-directed mutagenesis experiments demonstrated that functional binding sites for NF-κB and AP-1 in the hBD-2 promoter are required for induction of hBD-2 through *Escherichia coli* Nissle 1917. Treatment with the NF-κB inhibitor Helenalin, as well as with SP600125, a selective inhibitor of c-Jun N-terminal kinase, blocked hBD-2 induction by *Escherichia coli* Nissle 1917 in CaCO-2 cells. SB 202190, a specific p38 mitogen-activated protein kinase inhibitor, and PD 98059, a selective inhibitor of extracellular signal-regulated kinase 1/2, were ineffective. Thus probiotic bacteria could stimulate the intestinal innate defense through the upregulation of inducible antimicrobial peptides such as hBD-2. The induction of hBD-2 could contribute to an enhanced mucosal barrier to the luminal bacteria.

Pohjavuori et al., (2004) examined the role of probiotic bacteria in treatment of cow's milk allergy and IgE-associated dermatitis, by investigating the immunologic effects of *Lactobacillus* rhamnosus GG and a mixture of 4 bacterial species (MIX) [103]. A randomized, double-blind study design, was performed concomitantly with elimination diet and skin treatment. LGG, MIX, or placebo were given for 4 weeks to infants with suspected cow's milk allergy. After anti-CD3 (OKT3) and anti-CD28 stimulation of PBMCs, IFN-γ, IL-4, IL-5, and IL-12 levels were measured in culture supernatants by ELISA (figure 38). Intracellular IFN-γ, IL-4, and IL-5 production on CD4 lymphocytes was analyzed with fluorescence-activated cell sorting. Secretion of IFN-g by PBMCs before the treatment was significantly lower in infants with CMA ($p = 0.016$) and in infants with IgE-associated cow's milk allergy ($p = 0.003$) than in non-cow's milk allergy infants. Among the infants who received LGG, the level of secreted IFN-γ increased in those with cow's milk allergy ($p = 0.006$) and in those with IgE-associated dermatitis ($p = 0.017$) when compared with the placebo group. Secretion of IL-4 increased significantly in infants with cow's milk allergy in the MIX ($p = 0.034$) but not in the LGG group. Deficiency in IFN-γ response appeared to be related to cow's milk allergy. LGG raised IFN-γ production of PBMC in infants with cow's milk allergy and in infants with IgE-associated dermatitis and could thus provide beneficial T_H1 immunomodulatory signals. MIX, although containing LGG, appeared to modulate the immune responses differently.

In 2004, Chapat et al. examine whether *Lactobacillus casei* could affect antigen-specific T cell-mediated skin inflammation [104]. They used contact hypersensitivity to the hapten 2,4-dinitrofluorobenzene, a model of allergic contact dermatitis mediated by $CD8^+$ CTL and controlled by $CD4^+$ regulatory T cells. Daily oral administration of fermented milk containing *Lactobacillus casei* or *Lactobacillus casei* alone decreased skin inflammation by inhibiting the priming/expansion of hapten-specific IFN-γ-producing $CD8^+$ effector T cells. The down-regulatory effect of the probiotics required the presence of $CD4^+$ T cells, which control the size of the hapten-specific $CD8^+$ T cell pool primed by skin sensitization. *Lactobacillus casei* cell wall was as efficient as live *Lactobacillus casei* to regulate both the CHS response and the hapten-specific $CD8^+$ T cell response (figure 39), suggesting that cell wall components contribute to the immunomodulatory effect of *Lactobacillus casei*.

	INF-γ			IL-4			IL-5		
	n°	Before	After	n°	Before	After	n°	Before	After
CMA									
LGG	12	403	829	12	19	19	11	51	93
MIX	12	437	681	12	15	22	9	47	83
Placebo	9	836	579	9	21	18	8	56	45
Dermatitis (IgE)									
LGG	14	506	840	14	22	24	13	91	99
MIX	12	952	672	12	19	25	11	113	82
Placebo	12	1042	860	12	24	24	11	70	66
Dermatitis (IgE)									
LGG	24	780	859	24	19	19	23	51	79
MIX	21	492	663	21	18	23	18	69	73
Placebo	17	955	765	17	23	14	16	57	65

Figure 38. OKT3 and anti-CD3 induced cytokine secretion of PBMCs before and after treatment, medians (ranges), pg/mL [103].

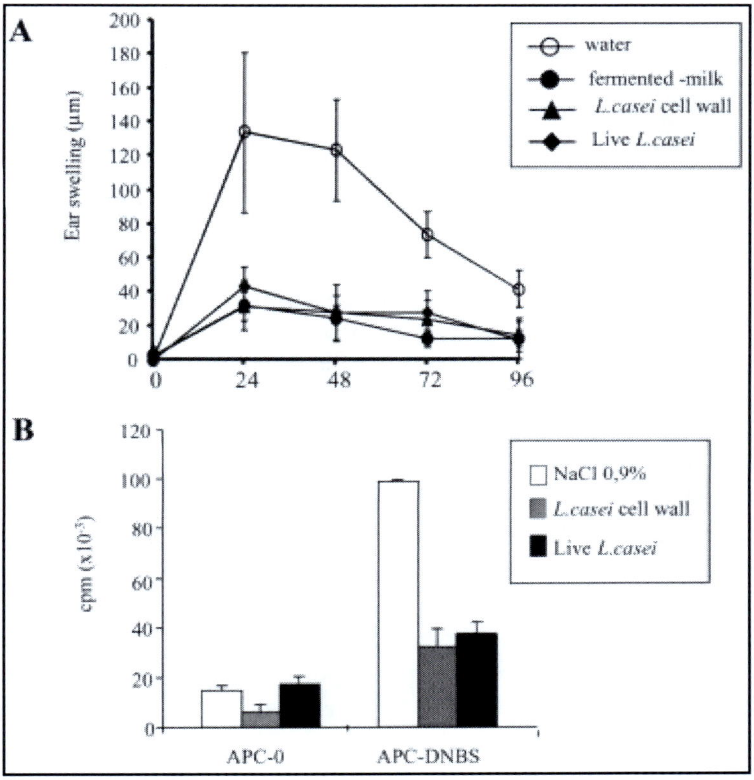

Figure 39. Inhibition of CHS to DNFB by *Lactobacillus casei*. Mice were treated daily from day 14 before skin sensitization with either 0.9% NaCl, fermented milk containing *Lactobacillus casei* DN-114 001, live *Lactobacillus casei* or *Lactobacillus casei* cell wall. (A) CHS was determined by ear swelling (lm) at various time points after challenge. The results are representative of five experiments. (B) The hapten-specific proliferative response of purified $CD8^+$ T cells was determined on day 5 after sensitization. $CD8^+$ T cells (2 x 10^5/well) were restimulated *in vitro* with mitomycin C-treated spleen cells (10^6/well) that were either incubated with medium alone (APC-0) or hapten derivatized (APC-DNBS). Hapten-specific proliferation was determined after 4 days of culture by [^{3}H]thymidine incorporation over the last 18 h of culture. The results are expressed as cpm (x 10^{-3}) [104].

The study provided the first evidence that oral administration of *Lactobacillus casei* can reduce antigen-specific skin inflammation by controlling the size of the $CD8^+$ effector pool.

Hart et al., (2004) assessed the effects of different probiotic bacteria on dendritic cell function. Human intestinal lamina propria mononuclear cells, whole blood, or an enriched blood dendritic cell population were cultured with cell wall components of the eight bacterial strains in the probiotic preparation VSL#3 (four lactobacilli, three bifidobacteria, and one streptococcal strains) [105]. Dendritic cells were identified and changes in dendritic cell maturation/co-stimulatory markers and cytokine production in response to probiotic bacteria were analyzed by multicolour flow cytometry, in addition to subsequent effects on T cell polarization. VSL#3 was a potent inducer of IL-10 (figure 40 A) by dendritic cells from blood and intestinal tissue, and inhibited generation of T_H1 cells. Individual strains within VSL#3 displayed distinct immunomodulatory effects on dendritic cells (figure 40 C); the most marked anti-inflammatory effects were produced by Bifidobacteria strains which up-regulated IL-10 production by dendritic cells (figure 40 B), decreased expression of the co-stimulatory molecule CD80, and decreased INF-γ production by T cells. VSL#3 diminished proinflammatory effects of LPS by decreasing LPS induced production of IL-12 while maintaining IL-10 production. Probiotic bacteria differ in their immunomodulatory activity and influence polarization of immune responses at the earliest stage of antigen presentation by dendritic cells.

Baillon et al., (2004) evaluated the viability of a probiotic strain of *Lactobacillus acidophilus* in a dry dog food, determined its ability to survive transit through the gastrointestinal tract and populate the colon, and assessed its effects on intestinal and systemic parameters [106]. Fifteen adult dogs were sequentially fed a dry control food for 2 weeks, the same food supplemented with more than 10^9 *Lactobacillus acidophilus* for 4 weeks, and the control food again for 2 weeks.

Fecal score was daily assessed, and fecal and blood samples were collected for enumeration of bacterial populations and measurement of hematologic variables. Recovery of *Lactobacillus acidophilus* from the supplemented food was 71% and 63% at the start and end of the study, respectively, indicating that the bacteria were able to survive manufacture and storage. The probiotic bacterium was detected in feces via ribotyping and RNA gene sequencing during the probiotic administration phase but not 2 weeks after cessation of administration. Administration of the probiotic-supplemented food was associated with increased numbers of fecal lactobacilli and decreased numbers of clostridial organisms. There were significant increases in RBCs, Hct, hemoglobin concentration, neutrophils, monocytes, and serum IgG concentration and reductions in RBC fragility and serum NO concentration.

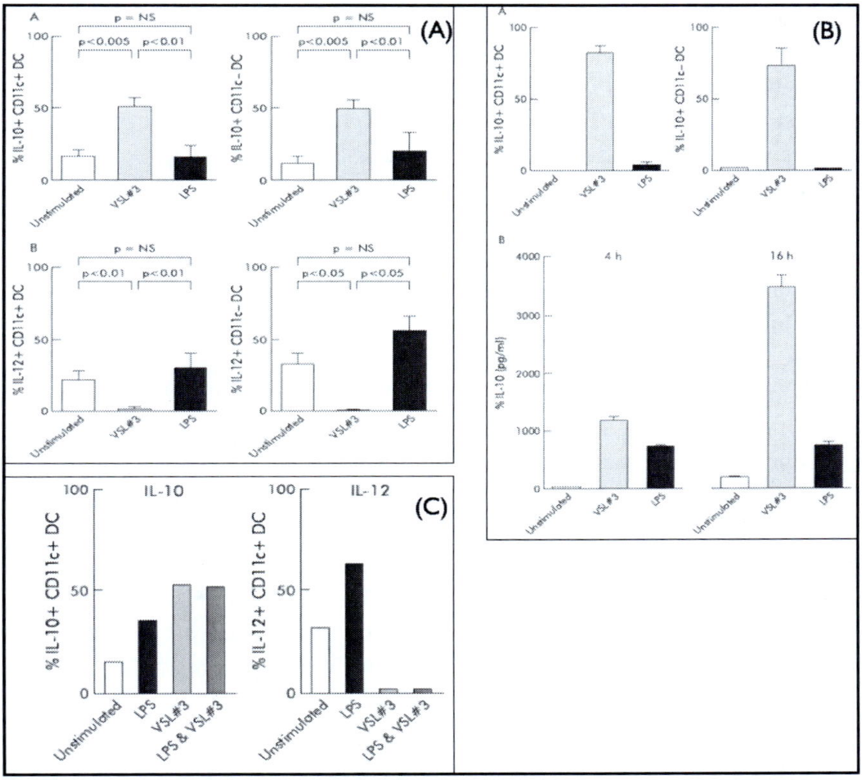

Figure 40. (A) Enhanced IL-10 production and decreased IL-12 production by blood DC in the presence of probiotic bacteria (VSL#3). Whole blood was incubated with control medium alone, VSL#3 probiotic bacteria (at a dose equivalent to 10^8 CFU/mL) or LPS (1 mg/mL) for four hours and (A) IL-10 production and (B) IL-12 production was assessed by intracellular cytokine staining. The proportion of cytokine staining $CD11c^+$ and CD11c- DC is shown (mean (SEM), n = 9-13); (b) enhanced IL-10 production by an enriched DC population in the presence of probiotic bacteria (VSL#3). (A) An enriched DC population (metrizamide separated) was cultured with control medium alone, VSL#3 probiotic bacteria (at a dose equivalent to 10^8 CFU/mL or LPS (1 mg/mL) for four h and the IL-10 production assessed by intracellular cytokine staining. (B) The concentration of IL-10 in the supernatants of enriched DC cultured with control medium, VSL#3 (at a dose equivalent to 10^8 CFU/mL) or LPS (1 mg/mL) was measured by ELISA at 4 h and 16 h; (c) Co-culture of VSL#3 with LPS decreased the proinflammatory effects of LPS. In whole blood, VSL#3 at 10^8 CFU/mL and LPS at 1 mg/mL were co-cultured and cytokine production assessed by intracellular staining. The proportion of CD11c+ DC producing IL-10 and IL-12 in these co-cultures is shown [105].

Data indicated that *Lactobacillus acidophilus* could be successfully incorporated into a dry dog food, survive transit through the canine gastrointestinal tract, and populate the colon and were associated with local and systemic changes. This probiotic bacterium could have the potential to enhance intestinal health and improve immune function in dogs.

The *in vitro* immune-stimulating effects on T_H1/T_H2 balance of more than 100 LAB strains were compared in splenocytes from ovalbumin-sensitized T_H2-polarized mice by Fujiwara et al., (2004) [107]. The *in vivo* anti-allergic ability of strain KW3110 was studied in the T_H2-polarized model by detecting serum IgE concentration, T_H1/T_H2 cytokine secretion from splenocytes, and the expression of co-stimulatory molecules on macrophages. *In vitro* studies from T_H2-polarized splenocytes, using IL-12 as a T_H1 parameter and IL-4 secretion as a T_H2 parameter revealed a wide variety of IL-12-inducing and IL-4-repressing activities, depending on the strain of LAB, not depending on the species. However, evaluation of individual strains *in vivo* revealed that after exposure to *Lactobacillus paracasei* KW3110 strain, the serum IgE elevation elicited by repeated OVA injection of mice was strongly inhibited. Cytokine secretion from splenocytes 20 weeks after KW3110 administration showed increased IL-12 and decreased IL-4 expression. Both CD40 and B7-1 expression on macrophages was upregulated by administration of KW3110. Improving the consequences of the T_H1/T_H2 imbalance by administration of LAB was dependent upon the LAB strain rather than the LAB species. Oral KW3110 administration in the mouse allergy model directed the T_H1/T_H2 balance toward Th1 through the maturation of APCs and inhibition of serum IgE elevation.

Gerbitz et al., in 2004, hypothesized that modifying the enteric flora of C57BL/6 and B6D2F1 mices, using the probiotic microorganism *Lactobacillus rhamnosus* GG, would ameliorate a Graft versus host disease (GVHD) [108]. They showed that oral administration of *Lactobacillus rhamnosus* GG before and after transplantation results in improved survival and reduced aGVHD. Furthermore, subculturing of mesenteric lymph node tissue revealed a reduced translocation of enteric bacteria. Our findings suggest that alteration of the intestinal microflora plays an important role in the initiation of experimental aGVHD.

Land et al., (2005) described 2 patients who received probiotic lactobacilli and subsequently developed bacteremia and sepsis attributable to *Lactobacillus* species [109]. Molecular DNA fingerprinting analysis showed that the *Lactobacillus* strain isolated from blood samples was indistinguishable from the probiotic strain ingested by the patients.

Sauter et al., (2005) investigated the influence of probiotics on mRNA and protein expression levels of cytokines in intestinal samples from dogs suffering from chronic enteropathies (CE) [110]. Duodenal samples of client-owned dogs with CE (group CE; n = 12) were collected during diagnostic endoscopy. Additional duodenal samples of gastrointestinally healthy dogs (group C; n = 4) from an unrelated study were available. Based on histopathological analyses, no pathological changes or only mild to moderate eosinophilic and/or lymphoplasmacytic duodenitis were diagnosed. Tissue samples were cultured: (1) with cell culture medium alone (negative control), (2) with a probiotic cocktail (PC), constituted of three *Lactobacillus* spp. from healthy canine fecal isolates, (3) with the individual strains of PC, and (4) with a placebo powder. Viability of intestinal tissue and probiotic bacteria before and after culture was evaluated. The mRNA abundance of interleukin (IL)-10, IL-12p40, IFN-γ, TNF-α, and TGF-1β was analyzed by RT-PCR. Protein concentrations of IFN-γ and IL-10 were measured in culture supernatant by ELISA. There was a loss of about 1 log CFU/mL of probiotic bacteria during the incubation period. Viability of tissue was maintained as confirmed by non-significant release of lactate dehydrogenase. In C, addition of PC increased IL-10 mRNA levels ($p < 0.1$). In CE, PC increased mRNA and protein levels of IL-10 ($p < 0.05$). On the mRNA level, the ratio of TNF-γ/IL-10, IFN-γ/IL-10, and IL-12p40/IL-10 decreased after addition of PC ($p < 0.05$) (figure 41). The results demonstrate favorable effects of PC on regulatory cytokines relative to inflammatory cytokines that might contribute to reduction of intestinal inflammation.

Parameter (mean value ± S.E.M.)	Culture medium				
	Medium	Probiotic cocktail	Placebo	*P*-value[a] (medium vs. PC)	*P*-value[a] (medium vs. placebo)
C ($n = 4$)					
IFN-γ (ng/mL)	28 ± 8	61 ± 47	4.5 ± 1.5	NS	NS
IL-10 (ng/mL)	20 ± 11	25 ± 2	13 ± 0.5	NS	NS
IFN-γ/IL-10	1.7 ± 0.5	2.4 ± 1.8	0.4 ± 0.1	NS	NS
CE ($n = 12$)					
IFN-γ (ng/mL)	154 ± 26	201 ± 39		NS	
IL-10 (ng/mL)	70 ± 22	161 ± 58		<0.05	
IFN-γ/IL-10	3.6 ± 0.9	2.6 ± 0.9		NS	

Figure 41. Protein analysis of culture supernatants. PC, probiotic cocktail; FRD, food responsive diarrhea; IBD, inflammatory bowel disease [110].

Lan et al., (2005) determine if primary murine colonic epithelial cells (CEC) respond to commensal bacteria and discriminate between different types of bacteria [111]. A novel CEC: bacteria co-culture system was used to compare the ability of the colonic commensal bacteria, *Bacteroides ovatus*, *Escherichia coli* (SLF) and *Lactobacillus rhamnosus* (LGG) to modulate production of different cytokines (n = 15) by primary CEC (figure 42).

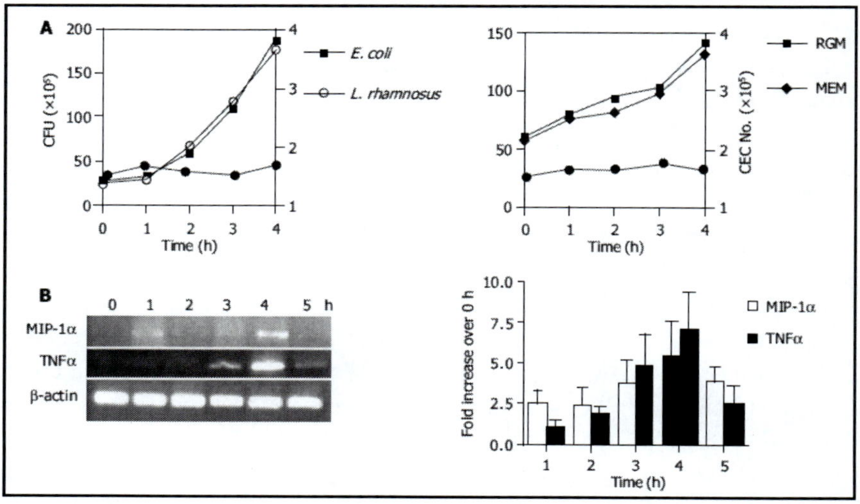

Figure 42. Bacterial growth (A) and kinetics of CEC cytokine gene expression (B) in CEC:bacteria co-cultures. A: Determination of bacterial (*Lactobacillus rhamnosus* and *Escherichia coli*) CFU by harvesting cells from CEC:bacteria co-cultures at hourly intervals up to 4 h by extensive washing of adherent CEC and plating serial dilutions onto agar plates and counting bacterial colonies 24 h later (left-hand panel). Bacteroides ovatus were cultured either alone under anaerobic conditions in RGM media or with CEC in 50 mL/L CO_2 and complete MEM (right-hand panel). CEC numbers (solid circle on both graphs) were determined by counting the number of cells recovered from co-cultures at the indicated times using a counting chamber; B: CEC cultured in the presence of *Escherichia coli* for up to 5 h. CECs were processed for RNA isolation and RT-PCR analysis using primers specific for the housekeeping gene β-actin, and MIP-1α and TNFα. Quantitative densitometry was carried out on EtBr-stained gels. Error bars indicate 95% confidence limits [111].

Antibody staining and flow cytometry were used to investigate Toll-like receptor (TLR) expression by CEC directly *ex vivo* and TLR responsiveness was determined by examining the ability of TLR ligands to influence CEC cytokine production. Primary CEC constitutively expressed functional TLR2 and TLR4. Cultured in complete medium alone, CEC secreted IL-6, MCP-1 and IP-10 the levels of which were significantly increased upon addition of the TLR ligands peptidoglycan (PGN) and lipopolysaccharide (LPS). Exposure to the commensal bacteria induced or upregulated different patterns of cytokine production and secretion. *Escherichia coli* induced production of MIP-1α/β and β defensin3 whereas *Bacteroides ovatus* and *Lactobacillus rhamnosus* exclusively induced MCP-1 and MIP-2α expression, respectively. TNF-α, RANTES and MEC were induced or up-regulated in response to some but not all of the bacteria whereas ENA78 and IP-10 were upregulated in response to all bacteria. Evidence of bacterial interference and suppression of cytokine production was obtained from mixed bacterial: CEC co-cultures. Probiotic *Lactobacillus* GG suppressed *Escherichia coli*, and *Bacteroides ovatus* induced cytokine mRNA accumulation and protein secretion. The ability of primary CEC to respond to and discriminate between different strains of commensal bacteria was assesd and a mechanism by which probiotic bacteria could exert anti-inflammatory effects *in vivo* was identified.

In 2005, Peña et al. investigated the role of TNF-α- inhibiting *Lactobacillus* clones with a pathogen-induced murine colitis model [112]. Murine-derived probiotic lactobacilli were selected *in vitro* for their ability to inhibit TNF-α secretion by *Helicobacter hepaticus*-stimulated macrophages. IL-10-deficient mice were treated with probiotic *Lactobacillus reuteri* in combination with *Lactobacillus paracasei* and then challenged with *Helicobacter hepaticus*. Ten weeks post-inoculation, the severity of typhlocolitis was assessed by histologic examination of the cecocolic region. Intestinal proinflammatory cytokine responses were evaluated by real-time quantitative reverse transcriptase PCR and immunoassays, and the quantities of intestinal *Helicobacter hepaticus* were evaluated by real-time PCR. Intestinal colonization by TNF-α-inhibitory lactobacilli reduced intestinal inflammation in Helicobacter hepaticus-challenged IL-10-deficient mice despite similar quantities of *Helicobacter hepaticus* in cocolonized animals. Proinflammatory colonic cytokine (TNF-α and IL-12) levels were lowered in *Lactobacillus*-treated animals. In this *Helicobacter hepaticus*-challenged IL-10-deficient murine colitis model, lactobacilli demonstrated probiotic effects by direct modulation of mucosal inflammatory responses.

Ukena et al., (2005) improved the understanding of the *Escherichia coli* Nissle 1917-host interaction by analyzing the gene expression pattern initiated by this probiotic in human intestinal epithelial cells [113]. Gene expression profiles of CaCO-2 cells treated with *Escherichia coli* Nissle 1917 were analyzed with microarrays. A second human intestinal cell line and also pieces of small intestine from BALB/c mice were used to confirm regulatory data of selected genes by real-time RT-PCR and cytometric bead array (CBA) to detect secretion of corresponding proteins. Whole genome expression analysis revealed 126 genes specifically regulated after treatment of confluent CaCO-2 cells with *Escherichia coli* Nissle 1917. Among others, expression of genes encoding the proinflammatory molecules monocyte chemoattractant protein-1 ligand 2 (MCP-1), macrophage inflammatory protein-2 alpha (MIP-2α) and macrophage inflammatory protein-2 beta (MIP-2β) was increased up to 10 fold. CaCO-2 cells cocultured with *Escherichia coli* Nissle 1917 also secreted high amounts of MCP-1 protein. Elevated levels of MCP-1 and MIP-2α mRNA could be confirmed with Lovo cells. MCP-1 gene expression was also up-regulated in mouse intestinal tissue. Thus, probiotic *Escherichia coli* Nissle 1917 specifically upregulated the expression of proinflammatory genes and proteins in both human and mouse intestinal epithelial cells.

Fernández et al., (2005) evaluated *Lactobacillus delbrueckii* subsp. *lactis* UO 004 for its use as a potential probiotic from a safety point of view [114]. The strain did not exhibit mucinolytic or other enzymatic activities that might be detrimental, such as those involving glycosidases (β-D-*glucosaminidase* or α-D-*galactosidase*) or arylamidases (factor Xa and quimotrypsin-like activities), frequently present in *Lactobacillus* strains isolated from patients with endocarditis, although it was able to express protein Ca and kallikrein-like activities. On the other hand, the presence of the strain did not interfere with the growth of certain species of normal intestinal microbiota, such as *Enterococcus fecalis*, *Escherichia coli*, *Bifidobacterium* bifidum or *Bacteroides fragilis*. Moreover, UO 004 was sensitive to antibiotics with transmissible resistance mechanisms in *Lactobacillus* such as chloramphenicol, erythromycin, tetracycline and vancomycin. In addition, *Lactobacillus delbrueckii* subsp. *lactis* UO 004 was not able to translocate towards the intestinal barrier of mice or produce changes in their activity or general health status.

Viljanen et al., in 2005, studied *in vivo* the immunologic effects of probiotics in infants with atopic eczema–dermatitis syndrome and cow's milk allergy [115]. Two hundred and thirty infants with atopic eczema–dermatitis syndrome and suspected cow's milk allergy received, concomitant with elimination diet, either LGG, a mixture of 4 probiotic strains (MIX), or placebo for 4 weeks. All available paired pretreatment and post-treatment plasma samples (n = 132) were analyzed for concentrations of IL-2, IL-4, IL-6, IL-10, TNF-α, IFN-γ, soluble intercellular adhesion molecule 1, soluble E-selectin, TGF-β1, TGF-β2, and C-reactive protein. In infants with IgE-associated AEDS, treatment with LGG induced higher C-reactive protein levels than in the placebo group (geometric mean, 0.83 mg/mL [95% CI, 0.56-0.81] vs 0.42 mg/mL [95% CI, 0.27-0.65]; ($p = 0.021$). Concomitantly, IL-6 levels increased after treatment with LGG ($p = 0.023$) but not with MIX or placebo. Soluble E-selectin levels were higher after probiotic than after placebo treatment in infants with IgE-mediated cow's milk allergy (LGG geometric mean, 86.7 ng/mL [95% CI, 75.2-100]; MIX geometric mean, 91.6 ng/mL [95% CI, 74.8-111.9]; and placebo geometric mean, 64.9 ng/mL [95% CI, 53-79.3]; analysis of covariance, $p = 0.035$; LGG *vs* placebo, $p = 0.023$; MIX *vs* placebo, $p = 0.020$). Use of MIX induced an increase in plasma IL-10 levels ($p = 0.016$). Probiotics induced systemically detectable inflammation downstage which might explain the clinical effects of probiotics in atopic eczema–dermatitis syndrome and cow's milk allergy.

Westendorf et al., (2005) tested the potential of *Escherichia coli* Nissle 1917 to serve as a safe carrier for targeted delivery of recombinant proteins to the intestinal mucosa [116]. In a well-defined and very sensitive immunological system, it was demonstrated that intestinal recombinant *Escherichia coli* Nissle 1917 (*Escherichia coli* Nissle-HA110-120) had no effect on migration, clonal expansion and activation status of specific $CD4^+$ T cells, neither in healthy mice nor in animals with acute colitis. Furthermore, recombinant *Escherichia coli* Nissle 1917 had no effect on the induction or breakdown of peripheral T-cell tolerance in an autoimmune environment. The excellent colonization properties of *Escherichia coli* Nissle 1917 render this strain an ideal candidate as carrier organism for gut-focused in situ synthesis of therapeutic molecules.

Olivares et al., (2006) described a randomized, double-blind, placebo-controlled human clinical trial to investigate the immune effects on healthy people of a fermented product containing two new probiotic strains, *Lactobacillus gasseri* CECT 5714 and *Lactobacillus coryniformis* CECT 5711, which was compared with another fermented product, a standard yogurt [117]. Consumption of either the new product or yogurt increased the proportion of phagocytic cells, including monocytes and neutrophils, as well as their phagocytic activity. However, combination of the product containing the strains *Lactobacillus gasseri* CECT 5714 and *Lactobacillus coryniformis* CECT 5711 also induced an increase in the proportion of natural killer (NK) cells and in IgA concentrations. The effects were higher after two weeks of treatment than after 4 weeks, which suggests regulation of the immune system. In addition, the new product enhanced immunity in the participants to a greater extent than did the control standard yogurt.

Vinderola et al., (2006) investigated the *in vivo* immunomodulating capacity of the exopolysaccharide produced by *Lactobacillus kefiranofaciens* (figure 43) by analyzing the profile of cytokines and immunoglobulins induced at the intestinal mucosa level, in the intestinal fluid and blood serum [118].

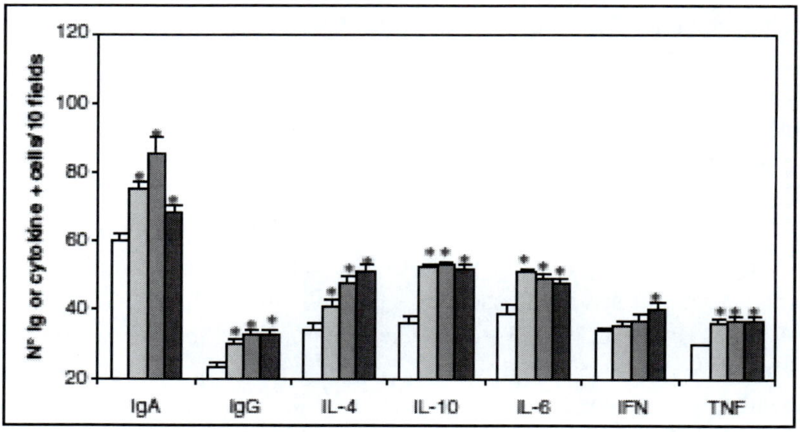

Figure 43. The immune effects of the exopolysaccharide kefiran on the large intestine lamina propria. In general, the effects observed here were more widespread than in the small intestine. All the immune parameters analyzed: the number of IgA^+, IgG^+, IL-4^+, IL-10^+, IL-6^+ and TNF-α^+ cells increased for all periods of administration, compared to control values. However, the number of IFN-γ^+ cells increased (significantly compared to control) only after 7 days of kefiran administration [118].

BALB/c mice received the exopolysaccharide produced by *Lactobacillus kefiranofaciens* for 2, 5 or 7 consecutive days. At the end of each period of administration, control and treated mice were sacrificed and the numbers of IgA^+ and IgG^+ cells were determined on histological slices of the small and large intestine by immunofluorescence. Cytokines (IL-4, IL-6, IL-10, IL-12, IFN-γ and TNF-α) were also determined in the gut lamina propria as well as in the intestinal fluid and blood serum. There was an increase of IgA^+ cells in the small and large intestine lamina propria, without change in the number of IgG^+ cells in the small intestine. The increase in the number of IgA^+ cells was not simultaneously accompanied by an enhance of the number of IL-4^+ cells in the small intestine. This finding would be in accordance with the fact that, in general, polysaccharide antigens elicit a T-independent immune response. For IL-10^+, IL-6^+ and IL-12^+ cells, the values found were slightly increased compared to control values, while IFN-γ^+ and TNF-α^+ cells did not change compared to control values. The effects observed on immunoglobulins and in all the cytokines assayed in the large intestine after kefiran administration were of greater magnitude than the ones observed in the small intestine lamina propria, which may be due to the saccharolytic action of the colonic microflora. In the intestinal fluid, only IL-4 and IL-12 increased compared to control values. In blood serum, all the cytokines assayed followed a pattern of production quite similar to the one found for them in the small intestine lamina propria. The exopolysaccharide induced a gut mucosal response and it was able to up and down regulate it for protective immunity, maintaining intestinal homeostasis, enhancing the IgA production at both the small and large intestine level and influencing the systemic immunity through the cytokines released to the circulating blood.

Kim et al., (2006) examined the mechanisms for the up-regulation of innate immune responses by a probiotic *Lactobacillus casei* ATCC27139, *in vivo* [119]. Using mouse models of systemic *Listeria monocytogenes* infection and MethA fibrosarcoma tumorigenesis (figure 44) in combination with BALB/c and SCID mice, it was found that parenteral administration of *Lactobacillus casei* ATCC27139 conferred a protective effect against *Listeria monocytogenes* infection and anti-tumor activity against MethA fibrosarcoma by activation of innate immunity, while *Lactobacillus casei* ATCC27139-J1R strains, which are J1 phage-resistant strains that have been selected from MNNG-treated clones, lacked these activities. Substantial differences between ATCC27139 and ATCC27139-J1R strains were observed in the capacity to induce innate cytokines such as TNF-α, IL-12, IL-18 and IFN-γ, and pathogen-associated molecular pattern receptors, TLR2 and Nod2, by spleen cells. In addition, although phosphorylation of NF-kB p65 in spleen was equally enhanced in the ATCC27139- and the ATCC27139-J1R-treated groups, phosphorylation of both p38 MAPK and MAPKAPK-2 was significantly induced only by ATCC27139.

Furthermore, inhibitors of NF-kB (sulfasalazine) and p38 MAPK (SB203580) significantly reduced cytokine production by the spleen cells of the mice treated with *Lactobacillus casei* ATCC27139, suggesting that both NF-kB and p38 MAPK signaling pathways played important roles in the augmentation of innate immunity by the probiotic *Lactobacillus casei*.

Cukovic et al., (2006) described a case of a 27-year-old male patient with a 6-month history of NOD2/CARD15-positive Crohn's disease with stenosis of the terminal ileum, previously treated with corticosteroids, that was admitted with fever, chills, diarrhea, fatigue and cachexia [120]. Diagnostic work-up revealed sepsis, an abscess in the right hepatic lobe, multiple abdominal collections and right-sided pleural effusion. Pus and blood cultures showed *Lactobacillus acidophilus*. The patient was discharged after 9 weeks in good general condition. Although extremely rarely, *Lactobacillus acidophilus* could cause hepatic abscess and sepsis. This could be explained by immunologic incompetence due to corticosteroid treatment of Crohn's disease.

Strain	Log_{10} viable Listeria in spleen (mean ± SD)				Tumor weight (g, mean ± SD)	
	BALB/c	SCID	C57BL6	C57BL6/beige	BALB/c	SCID
Saline	5.78 ± 0.14	5.54 ± 0.22	5.88 ± 0.20	5.90 ± 0.18	3.27 ± 0.56	2.17 ± 0.46
27139	3.79 ± 0.16*	3.63 ± 0.23*	3.63 ± 0.18*	4.19 ± 0.34*	0.67 ± 0.27 (79)[c],*	0.47 ± 0.08 (78)*
27139-J1R	6.16 ± 0.13	5.43 ± 0.22	5.70 ± 0.37	6.32 ± 0.02	3.06 ± 0.72 (7)	1.66 ± 0.35 (24)

Figure 44. Requirement for innate immunity in augmentation of host resistance against *Listeria*[a] and MethA fibrosarcoma[b] by probiotic lactobacilli [119].

Duncker et al., (2006) studied the effect of the probiotic strain *Escherichia coli* Nissle 1917 (EcN) administered orally to young pigs at two concentrations (10^9 and 10^{11} CFU/d for 21 days) on the gut-associated lymphatic tissue [121]. This probiotic strain was shown recently to reduce recurrence of inflammation in ulcerative colitis patients. The number and distribution of intestinal immune cells (granulocytes, mast cells, $CD4^+$, $CD8^+$, $CD25^+$, IgA^+ lymphocytes) were highly different between small intestinal and colon segments in all groups, but were not influenced by EcN, except high dose EcN fed pigs (10^{11} CFU/d) showing an increase in mucosal $CD8^+$ cells in the ascending colon. The mRNA analysis revealed no changes associated with EcN feeding. Thus according to analyses EcN had only minor effects on the distribution of mucosal immune cells in the gut of healthy individuals. The well-established preventive effects of EcN could therefore be related to other mechanisms than simple modulation of immune cell distribution.

Marshall-Jones et al., (2006) evaluated the effect of dietary supplementation with the probiotic strain *Lactobacillus acidophilus* DSM13241 in healthy adult cats [122]. Fifteen adult cats were fed a nutritionally complete dry food for 5 weeks. Fecal character was assessed daily, and a single fecal sample and 3-mL blood sample were collected for bacterial enumeration and hematologic analysis, respectively. Cats were then fed the same diet supplemented with *Lactobacillus acidophilus* DSM13241 (2 x 10^8 CFU/d) for 4.5 weeks. Repeat fecal and hematologic measurements were taken prior to the return to control diet for a 4-week period. The probiotic species was recovered from the stools, demonstrating survival through the feline gastrointestinal tract. Probiotic supplementation was associated with increased numbers of beneficial *Lactobacillus* and *Lactobacillus acidophilus* groups in feces and decreased numbers of *Clostridium* spp and *Enterococcus faecalis*, indicating an altered bacterial balance in the gastrointestinal tract microflora. Fecal pH was also decreased suggesting a colonic environment selective for the beneficial lactic acid bacterial population. Systemic and immunomodulatory effects were associated with administration of *Lactobacillus acidophilus* DSM13241 including altered cell numbers within WBC subsets and enhanced phagocytic capacity in the peripheral granulocyte population. In addition, plasma endotoxin concentrations were decreased during probiotic feeding, and RBCs had a decreased susceptibility to osmotic pressure. Probiotic strain *Lactobacillus acidophilus* DSM13241 fed at 2 x 10^8 CFU/d could alter the balance of gastrointestinal microflora in healthy cats. Furthermore, administration of this probiotic results in beneficial systemic and immunomodulatory effects in cats.

Voltan et al., (2007) tested the relevance of the autoaggregation phenotype on the intestinal persistence of lactobacilli and their ability to modulate the host's innate immune responses [123]. After 14 days of diet supplementation, the aggregating strain *Lactobacillus crispatus* M247 but not aggregation-deficient isogenic mutant MU5 was recovered from the feces and colonic mucosa of mice. This observation was confirmed by strain-specific PCR amplification and by *Lactobacillus*-specific denaturing gradient gel electrophoresis analysis. Indeed, *Lactobacillus crispatus* M247 increased Toll-like receptor 2 (TLR2) mRNA levels, while it reduced TLR4 mRNA and protein levels in the colonic mucosa, whereas MU5 was ineffective. In colonic epithelial cells (CMT-93 cells) *Lactobacillus crispatus* M247 but not MU5 induced time-dependent extracellular signal-regulated kinase-1 (ERK1) tyrosine phosphorylation and TLR modulation, which were abolished in the presence of PD98059 (an ERK1 inhibitor). To assess the functional relevance of probiotic-induced TLR modulation, they determined the consequences of *Lactobacillus crispatus* preexposure on TLR4 (lipopolysaccharide) and TLR2 [Pam3Cys-Ser-(Lys)4] ligand-mediated effects in intestinal epithelial cells. Preexposure to *Lactobacillus crispatus* M247 blunted LPS-induced IL-6 release and inhibition of CMT-93 migration over a wound edge, whereas it enhanced TLR2-mediated IL-10 up-regulation. The aggregation phenotype was required for *Lactobacillus crispatus* persistence in the colon and for modulation of TLR2/TLR4 expression through an ERK-dependent pathway. The aggregation phenotype in *Lactobacillus crispatus* M247 could be required to temper epithelial cell responsiveness to bacterial endotoxins, which thus affects the evolution of intestinal inflammatory processes.

Galdeano et al., (2006) studied the influence of the probiotic bacterium *Lactobacillus casei* in the expression of receptors involved in the innate immune response [124]. This bacterium was orally administered to BALB/c mice which after were sacrificed; the small intestine and intestinal fluids were collected to measure secretory immunoglobulin A (IgA) specific for *Lactobacillus casei*. Mononuclear cells from Peyer's patches were isolated to determine the CD-206 and TLR-2 receptors. In histological slices we determined the number of IgA^+, $CD4^+$, $CD8^+$, and $CD3^+$ cells and two cytokines (IL-5 and IL-6). CD-206 and TLR-2 increased with respect to the untreated control. An increase in the T population or in the IL-5-positive cells was not observed. IgA^+ cells and IL-6-producing cells increased after 7 days of *Lactobacillus casei* administration. Specific antibodies against *Lactobacillus casei* were not found. The main immune cells activated after oral *Lactobacillus casei* administration were those of the innate immune response, with an increase in the specific markers of these cells (CD-206 and TLR-2), with no modification in the number of T cells.

Castellazzi et al., (2007) evaluated the *in vitro* activity of *Lactobacillus paracasei* (I 1688), *Lactobacillus salivarius* (I 1794), and a commercial mix of the two (PSMIX, Proge Farm), on immune cells from healthy individuals [125]. Two probiotic strains, *Lactobacillus salivarius* (I 1794; Proge Farm, Italy) and *Lactobacillus paracasei* (I 1688; Proge Farm, Italy), which are contained in the functional food "Enterobacilli", were evaluated for their ability to stimulate peripheral blood mononuclear cells and modulate surface phenotype and cytokine production. All subjects responded to the bacteria, with different levels of response. The cell populations that showed a significant percent increase were $CD4^+/CD25^+$ cells (T-helper activated regulatory cells), $CD8^+/CD25^+$ (T-suppressor/cytotoxic activated cells), and $CD16^+/CD56^+$ (NK cells) ($p < 0.05$). IL-12 and IFN-γ *in vitro* production significantly increased with exposure to probiotics ($p < 0.05$ for both). This study provided the evidence that *Lactobacillus paracasei* and *Lactobacillus salivarius* were capable of inducing a specific immune response that could be useful in the clinical setting for improving innate and adaptive immune responses.

Scharek et al., (2007), in a feeding trial, fed sows and piglets with the probiotic bacterium *Bacillus cereus* var. *toyoi* as a feed additive, and examined the effects on immune cell populations [126]. The development of the gut immune system was determined for piglets at the ages of 14, 28, 35 and 56 days post partum. Tissue samples of the jejunum and the continuous Peyer's patch were used for enumeration of intraepithelial lymphocyte populations by fluorescence activated flow cytometry and fluorescence microscopy. Both independent methods of investigation led to similar results: the population of intraepithelial $CD8^+$ T cells was significantly enhanced in the probiotic group piglets ($p \leq 0.05$), and the numbers of $\gamma\delta$ T cells tended to be higher in the intestinal epithelium ($p < 0.1$) at the time of weaning (day 28). Lamina propria lymphocytes were also influenced by the treatment. Application of *Bacillus cereus* var. *toyoi* resulted in significantly more $CD25^+$ lymphocytes and gd T cells in the probiotic group post-weaning (figure 45). The occurrence of pathogenic *Escherichia coli* serogroups was also less frequent in the feces of piglets from the probiotic group. The finding that the $CD8^+$ T cell population in the intestinal mucosa showed changes on day 28 indicated that the influence of *Bacillus cereus* var. *toyoi* supplementation on the intestinal immune system started before weaning, an observation supported by changes in the intestinal microflora observed during the suckling-period.

Age (days)	CD21		CD4		CD8	
	Control	Probiotic	Control	Probiotic	Control	Probiotic
14	69.8 ± 11.5	45.9 ± 63.7	4.6 ± 2.8	3.8 ± 5.7	6.3 ± 2.5	3.5 ± 1.1
28	78.1 ± 21.9	42.8 ± 29.6	2.9 ± 1.9	4.5 ± 3.9	5.2 ± 0.6	5.5 ± 1.6
35	90.1 ± 3.6	82.6 ± 17.6	1.7 ± 1.2	1.1 ± 0.4	2.9 ± 1.2	3.1 ± 1.0
56	73.1 ± 15.2	84.0 ± 11.8	2.7 ± 2.1	3.0 ± 1.7	10.1 ± 10.9	7.0 ± 3.9

Figure 45. Mean percentages (arithmetic mean) and standard deviations of CD4+ and CD8+ cells counted in suspension from continuous ileal PPs [126].

(A)							
Serological parameters	Group 1 (n = 16)			Group 2 (n = 11)			*p* value
	Baseline	Day 7	*p* value	Baseline	Day 7	*p* value	
CRP (mg/dL)	2.40±0.78	0.30±0.21	<0.05	1.20±0.50	0.35±0.20	NS	NS
IgA (mg/dL)	40.0±5.16	55.25±7.0	<0.05	31.14±4.09	35.80±6.18	NS	NS
IgG (mg/dL)	804.87±83.12	908.06±78.39	NS	671.54±66.55	719.45±72.2	NS	NS
IgM (mg/dL)	117.80±14.64	134.01±9.5	NS	87.18±13.18	118.7±21.88	NS	NS

B							
Immunological parameters	Group 1 (n = 16)			Group 2 (n = 11)			*p* value
	Baseline	Day 7	*p* value	Baseline	Day 7	*p* value	
CD3 (%)	61.31±3.14	65.23±3.36	NS	62.01±3.47	66.60±3.77	NS	NS
CD4 (%)	37.03±2.55	38.43±2.22	NS	37.87±3.77	42.20±2.75	NS	NS
CD8 (%)	18.57±1.65	24.41±1.56	<0.05	20.65±2.77	22.18±2.61	NS	<0.05
CD4/CD8 ratio	2.18±0.20	1.72±0.14	<0.05	2.17±0.36	2.14±0.24	NS	NS
Nk cells (%)	14.01±1.60	15.17±1.78	NS	12.32±2.23	9.99±1.38	NS	NS
TNF-α (ng/mL)	7.06±3.29	3.78±2.62	NS	5.07±2.34	4.01±1.98	NS	NS
TGF-β (pg/mL)	4.03±0.81	4.42±0.72	NS	3.55±0.64	3.60±0.86	NS	NS
C3 (mg/dL)	125.96±7.65	112.20±6.45	NS	136.06±12.4	120.0±12.09	NS	NS
C4 (mg/dL)	24.99±2.83	17.16±2.13	<0.05	26.12±2.96	20.14±2.41	<0.05	NS

C					
Lymphocytes	Group 1 (n = 6)		Group 2 (n = 6)		*p* value
	Baseline	Day 7	Baseline	Day 7	
CD4 (%)	22.85±11.92	15.30±9.02	22.40±5.43	17.16±6.60	0.041
CD8 (%)	13.91±2.53	21.93±6.29	19.36±4.38	21.93±6.29	0.015
CD4/CD8 ratio	2.73±0.76	1.83±0.53	1.72±0.65	2.09±0.64	0.015

Figure 46. (A) Comparison of serological parameters between children (n = 27) with acute diarrhoea on admission to the paediatric clinic (baseline) and after 7 days of treatment with either 250 mg *Saccharomyces boulardii* dissolved in 5 mL of water orally twice daily (group 1) or placebo (group 2); (B) Comparison of immunological parameters between children (n = 27) with acute diarrhoea on admission to the paediatric clinic (baseline) and after 7 days of treatment with either 250 mg *Saccharomyces boulardii* dissolved in 5 mL of water orally twice daily (group 1) or placebo (group 2); (C) Comparison of the proportion of CD4 and CD8 lymphocytes and the CD4/CD8 ratios between infants who were breast-fed for more than 4 months in both treatment groups (250 mg *Saccharomyces boulardii* dissolved in 5 mL of water orally twice daily for 7 days [group 1] or placebo [group 2]). (Modified from Ozkan et al., 2007) [128].

The results suggest that feeding of *Bacillus cereus* var. *toyoi* to sows may result in beneficial effects on piglet health status independent of their feed supplementation.

Nerstedt et al., (2007) used gene array technology to map the pattern of changes in the global gene expression profile of the host caused by *Lactobacillus* administration [127]. Microarrays were applied to comparatively characterize differences in gene transcription in the distal ileum of normal microflora (NMF) and germ-free (GF) mice evoked by oral administration of two *Lactobacillus* strains used in fermented dairy products today - *Lactobacillus paracasei* subsp. *paracasei* F19 (L. F19) or *Lactobacillus acidophilus* NCFB 1748. Feeding either of the two strains caused very similar effects on the transcriptional profile of the host. Both L. F19 and *Lactobacillus acidophilus* NCFB 1748 evoked a complex response in the gut, reflected by differential regulation of a number of genes involved in essential physiological functions such as immune response, regulation of energy homeostasis and host defence. Notably, the changes in intestinal gene expression caused by *Lactobacillus* were different in the mice raised under GF *vs.* NMF conditions, underlying the complex and dynamic nature of the host-commensal relationship. Differential expression of an array of genes evoked novel hypothesis of possible interactions between the probiotic bacteria and the host organism and warrants further studies to evaluate the functional significance of these transcriptional changes on the metabolic profile of the host.

Ozkan et al., (2007) evaluated the effect of *Saccharomyces boulardii* administration in otherwise healthy children aged between 6 months and 10 years who were admitted for acute diarrhoea (15 males, 12 females) [128]. Patients were randomized into two groups: group 1 (n = 16) received 250 mg *Saccharomyces boulardii* dissolved in 5 mL of water orally twice daily for 7 days and group 2 (n = 11) received placebo. Clinical and laboratory assessments were performed on admission and on day 7 of follow-up. Both groups experienced reduced daily stool frequency, the decrease being significantly greater in group 1 on days 3 and 4 compared with group 2. Group 1 demonstrated significant increases in serum immunoglobulin A and decreases in C-reactive protein levels on day 7 (figure 46 A-B). The percentage of CD8 lymphocytes on day 7 was significantly higher in group 1 than group 2 (figure 46 C). The efficacy of *Saccharomyces boulardii* in paediatric acute gastroenteritis and the findings suggested that *Saccharomyces boulardii* treatment enhanced the immune response.

Niers et al., (2007) investigated the capacity of probiotic bacteria to interact with neonatal dendritic cells (DC) and studied the ensuing T cell polarizing effect [129]. Immature DC were generated from cord blood-derived monocytes and maturation was induced by maturation factors (MF), lipopolysaccharide (LPS) plus MF and *Bifidobacterium bifidum*, *Bifidobacterium infantis*, *Lactobacillus salivarius*, *Lactococcus lactis* alone or combined with MF.

After 12 days of co-culture with DC and *Staphylococcus aureus* enterotoxin B (SEB) as antigenic stimulus, cytokine production by autologous T cells was determined by intracellular cytokine staining (figure 47). Additionally, cells were stimulated with CD3 and CD28 monoclonal antibodies and cytokines were measured in supernatants by multiplex assay. The probiotic strains induced partial maturation of DC.

Full maturation of DC was induced for all strains tested when MF was added. The percentage of IL-4 producing T cells was lower in T cell cultures stimulated with *Bifidobacterium bifidum* matured DC compared to MF and LPS matured DC, which coincided with a higher percentage of IFN-γ -producing T cells. Furthermore, T cells stimulated by *Bifidobacterium bifidum* matured DC produced significantly more IL-10 compared to MF matured DC. Selected species of the *Bifidobacterium* genus prime *in vitro* cultured neonatal DC to polarize T cell responses and could therefore be candidates to use in primary prevention of allergic diseases.

Foligné et al., (2007) investigated the correlation between the *in vitro* immune profile of probiotic strains and their ability to prevent experimental colitis in mice [130]. *In vitro* immunomodulation was assessed by measuring IL-12p70, IL-10, TNF-α and IFN-γ release by human peripheral blood mononuclear cells (PBMCs) after 24 hours stimulation with 13 live bacterial strains (figure 48).

A murine model of acute TNBS-colitis was next used to evaluate the prophylactic protective capacity of the same set of strains. A strain-specific *in vivo* protection was observed. The strains displaying an *in vitro* capacity to induce higher levels of the anti-inflammatory cytokine IL-10 and lower levels of the inflammatory cytokine IL-12, offered the best protection in the *in vivo* colitis model.

In contrast, strains leading to a low IL-10/IL-12 cytokine ratio could not significantly attenuate colitis symptoms. It could be possible to predict the *in vivo* protective capacity of the studied LAB based on the cytokine profile. The PBMC-based assay used could thus serve as a useful primary indicator to narrow down the number of candidate strains to be tested in murine models for their anti-inflammatory potential.

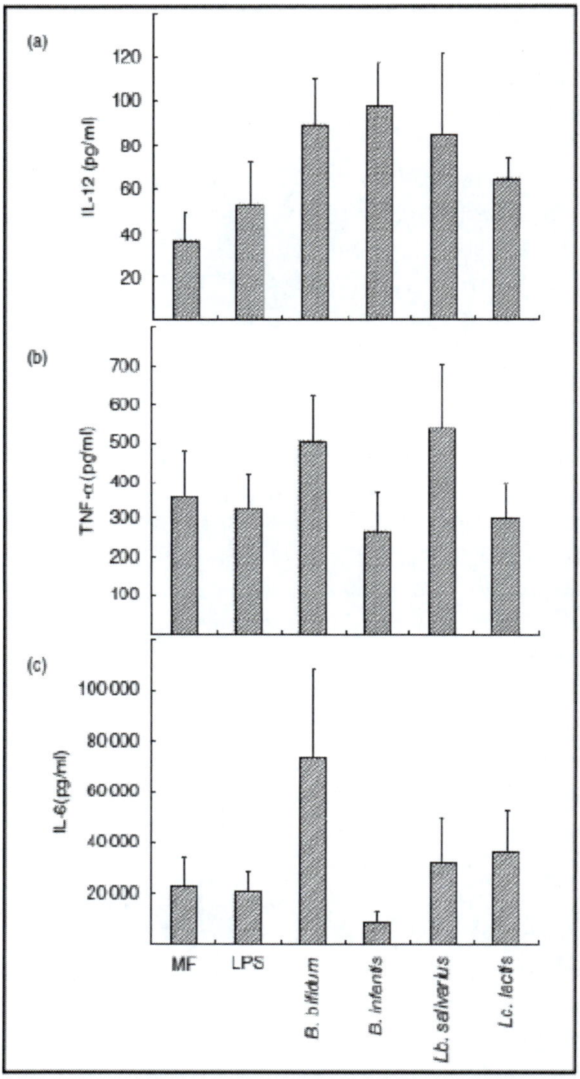

Figure 47. Cytokine production by dendritic cells (DC) upon exposure to probiotic bacteria. Mature DC (day 8 of culture) were stimulated overnight with CD40-ligand transfected J558 cells to induce production of cytokines. IL-12, TNF-α and IL-6 were measured in supernatants by multiplex assay. Data represent mean cytokine production standard error of the mean (n = 7). Compared to maturation factors (MF) matured DC or lipopolysaccharide (LPS)-matured DC the presence of the four tested strains during maturation did not have any statistically significant effect ($p > 0{\cdot}05$) on the production of IL-12, TNF-a and IL-6 by DC. Mann–Whitney U-test was used [129].

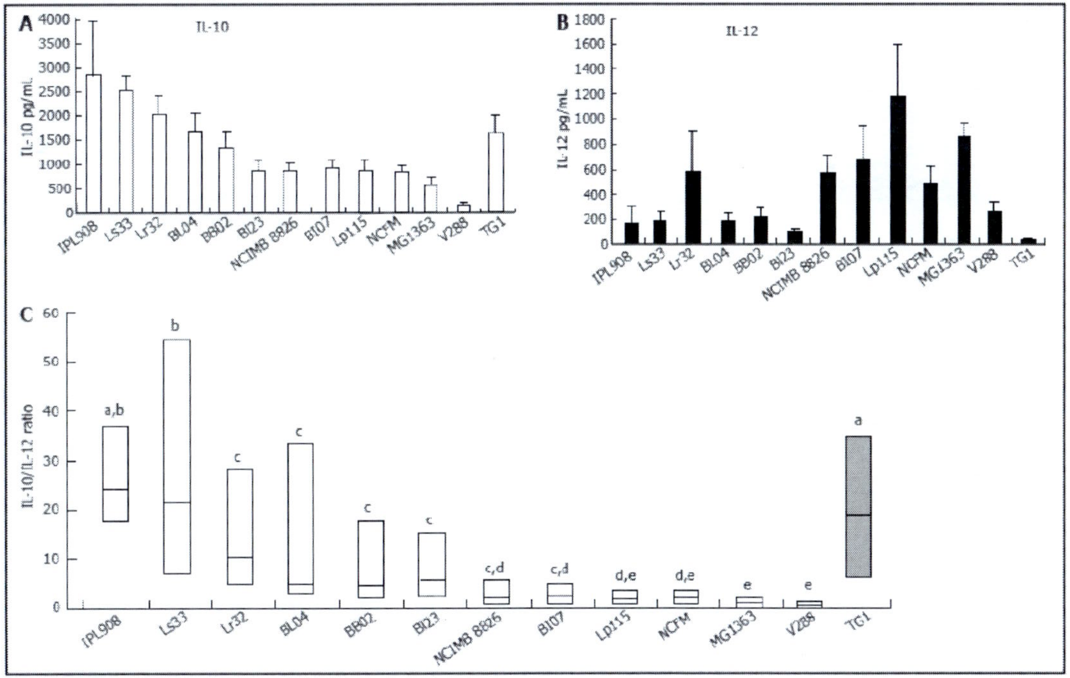

Figure 48. Strain-specific patterns of IL-10 (A) and IL-12p70 (B) release for various bacterial strains and IL-10/IL-12 ratios (C) for 6 to 12 independent healthy donors. Bars represent the mean ± SE values in pg/mL for 6 to 12 independent healthy donors. Ranked box and whisker plots show the median values and first to third quartiles in boxes. Different letters indicate significant differences according to Mann-Whitney U test ($p < 0.05$) [130].

Vinderola et al., (2007) studied in mice the impact of the oral administration of the supernatant of milk fermented by *Lactobacillus helveticus* R389 on the gut physiology by measuring parameters such as calcium channels and E-cadherin expression, the activation of the biological signal calcineurin and mast and goblet cells, as a way to determine some mechanisms involved in the immunomodulating effects of the milk fermentation products, observed in previous studies [131]. The impact of the supernatant of milk fermented by *Lactobacillus helveticus* R389 at pH6-controlled on the expression of calcineurin and on the reinforcement of the ephitelial barrier was analyzed, measuring parameters such as calcium channels and E-cadherin expression and in the reinforcement of the non-specific immunity determining mast cells and goblet cells associated to the gut. An enhanced expression of TRPV6 channels in the duodenum was observed, indicating an improved capacity for dietary Ca_2^+ uptake. An enhanced expression of calcineurin in the small intestine was demonstrated, able to upregulate immune parameters such as IL-2 and TNF production, with an increase in the number of these cytokines secreting cells. An increase in the number of mucosal mast cells and goblet cells was demonstrated, which would mean an improved state of mucosal surveillance at sites of infection. The oral administration of the supernatant of milk fermented by *Lactobacillus helveticus* R389 enhanced the gut mucosal immunity by improving the mechanisms that reinforce the epithelial and non-specific barriers and the gut functioning at sites of infection, with an improvement in the expression of the enzyme calcineurin, an important signal in the network that activates the gut immune system.

Chuang et al., (2007) investigated the effects on immune functions, via stimulating splenocytes, with three heat-killed Lactobacillus strains and their effect on mouse dendritic cells (DCs) and the related T cell responses [132]. The results showed that these *Lactobacillus* strains were able to stimulate cell proliferation and IL-10, IL-12 p70, and IFN-γ production but not TGF-β in splenocytes (figure 49). In addition, these heat-killed *Lactobacillus* strains also stimulated high-level secretion of IL-12 p70 in DCs and switched T cells to T helper T_H1 immune responses, as evidenced by the elevated secretion of IFN-γ but not IL-5, IL-13, and TGF-β (figure 50). Lactobacilli played a potentially important role in modulating immune responses and allergic reactions.

Lara-Villoslada et al., (2007) evaluated the oral toxicity of the recently isolated probiotic bacteria *Lactobacillus coryniformis* CECT5711 and *Lactobacillus gasseri* CECT5714 [133]. Enzymatic activity and antibiotic resistance profile were evaluated *in vitro*. Then, the oral toxicity was analyzed by an *in vivo* experiment using 20 Balb/C mice, which were orally treated with CECT5711 or CECT5714 (10^{10} CFU mouse/day) during 30 days. Results showed that CECT5711 and CECT5714 had no deleterious enzymatic activities and present intrinsic antibiotic resistance profile. Administration of both strains to mice had no adverse effects on body weight or food intake. No bacteraemia was present in liver or spleen and there was no treatment-associated bacterial translocation to these tissues. Liver glutathione content as well as plasma malondialdehide concentration were not statistically different in probiotic-treated mice when compared with control mice. Probiotic treatment did not cause changes in the biochemical and haematological parameters analyzed. Results suggested that strains CECT5711 and CECT5714 were nonpathogenic and likely to be safe for human consumption.

Matsuzaki et al., (2007) investigated the effects of *Lactobacillus casei* (LcS) in autoimmune disease models using nonobese diabetic (NOD) mice [134]. NOD mice spontaneously develop diabetes resembling human insulin-dependent diabetes mellitus (IDDM). These mice showed progressive lymphocytic infiltration of autoreactive $CD8^+$ T cells into the islets of Langerhans, and they have been recognized as a model of autoimmune disease, with cytoplasmic antibodies to islet cells appearing in their serum during the development of insulitis. From the age of 4 weeks, female NOD mice were fed a diet containing 0.05% by weight of LcS, and the onset of diabetes was recorded thereafter.

The incidence of diabetes in the control group was significantly higher than that in the LcS-treated group, and pathological analysis of the LcS-treated group revealed strong inhibition of the disappearance of insulin-secreting β-cells in Langerhans islets (figure 51). Moreover, the proportion of $CD8^+$ T cells among spleen cells was decreased in the LcS treated group, suggesting the inhibition of autoreactive T cells. Although the detailed mechanism of action of LcS in this model was not been clarified, it was postulated that LcS could alter the imbalance of T_H1/T_H2 cytokine production, which is thought to be the cause of the onset of IDDM.

Arslanoglu et al., (2007) studied the effect of a mixture of four probiotic bacterial strains along with prebiotic galacto-oligosaccharides in preventing allergic diseases [135]. One thousand two hundred twenty-three pregnant women carrying high risk children were randomized to use a probiotic preparation or a placebo for 2 to 4 weeks before delivery. Their infants received the same probiotics plus galacto-oligosaccharides (n = 461) or a placebo (n = 464) for 6 months.

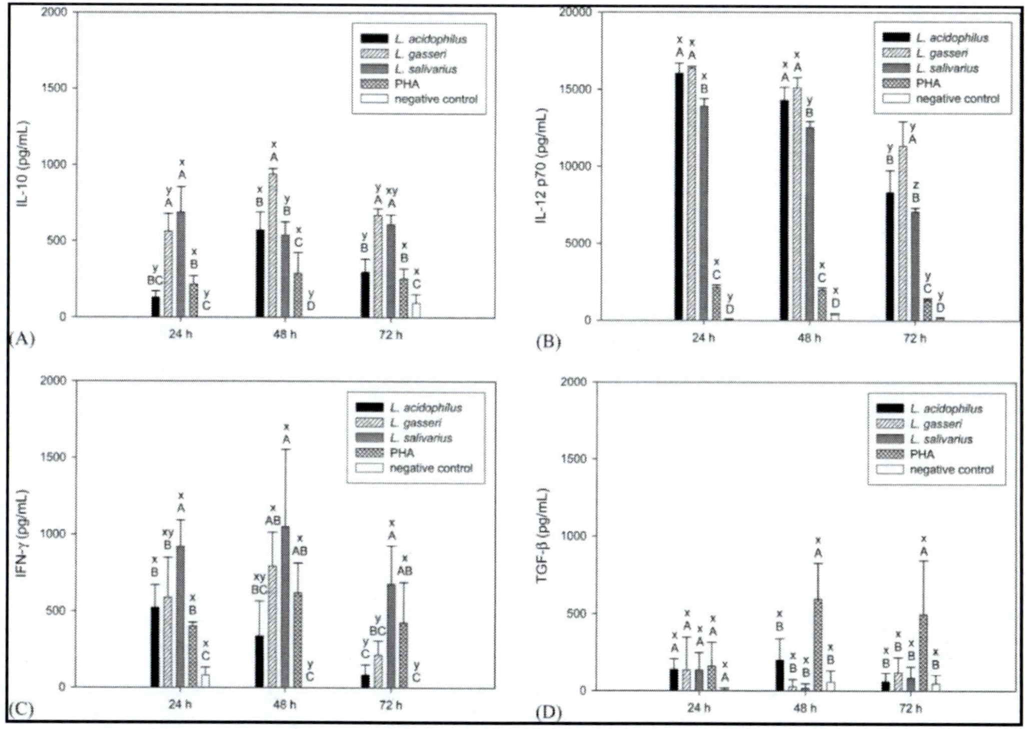

Figure 49. Production of IL-10 (A), IL-12 p70 (B), IFN-γ (C), and TGF-β (D) in splenocytes treated with heat-killed *Lactobacillus* strains. Splenocytes were treated with heat-killed lactobacilli, phytohemagglutinin (PHA), or no stimulants. Data are presented as mean ± SD (n = 3). Bars with different letters A-D are significantly different among different treatments ($p < 0.05$). Bars with different letters x-z are significantly different among different time points ($p < 0.05$) [132].

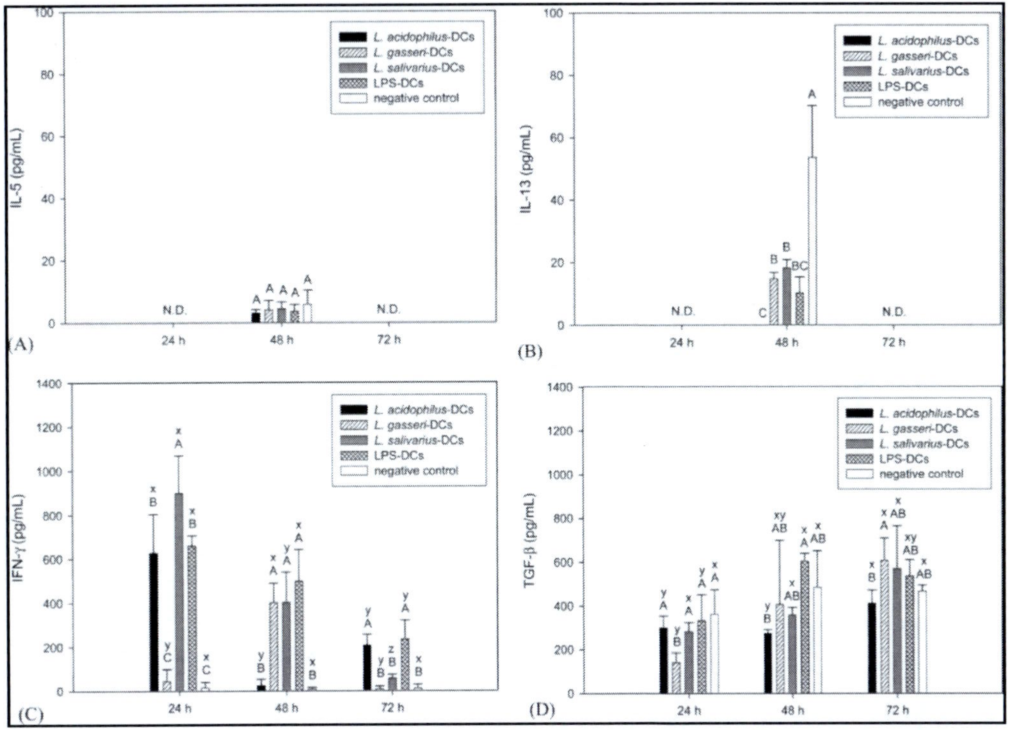

Figure 50. Production of IL-5 (A), IL-13 (B), IFN-γ (C), and TGF-β (D) in T cells induced by DCs treated with heat-killed *Lactobacillus* strains. DCs were treated with heat-killed *Lactobacillus* strains, LPS, or no stimulants for 24, 48, and 72 h and then co-cultured with T cells. ND: not detectable. Bars with different letters A-C are significantly different among different treatments ($p < 0.05$). Bars with different letters x-z are significantly different among different time points ($p < 0.05$) [132].

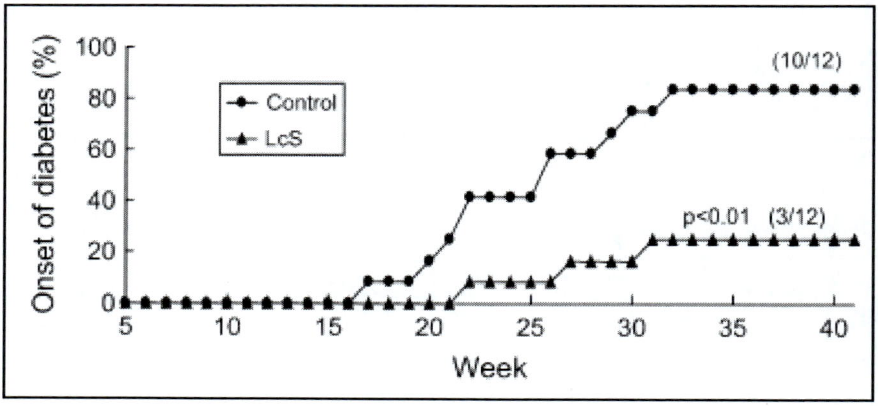

Figure 51. Effect of LcS on the onset of diabetes in female NOD mice. 4weeks-old NOD mice were fed a control diet (n = 12) (●) or a diet containing 0.05% LcS (n = 12) (▲), and the onset of diabetes was monitored for 40 weeks. Numbers in parentheses indicate the number of mice with onset-tested animals [134].

At two years the cumulative incidence of allergic diseases (food allergy, eczema, asthma, and allergic rhinitis) and IgE sensitization (positive skin prick test response or serum antigen-specific IgE level > 0.7 kU/L) were evaluated. Fecal bacteria were analyzed during treatment and at age 2 years. Probiotic treatment compared with placebo showed no effect on the cumulative incidence of allergic diseases but tended to reduce IgE-associated (atopic) diseases (odds ratio [OR], 0.71; 95% CI, 0.50-1.00; $p = 0.052$). Probiotic treatment reduced eczema (OR, 0.74; 95% CI, 0.55-0.98; $p = 0.035$) and atopic eczema (OR, 0.66; 95% CI, 0.46-0.95; $p = 0.025$). Lactobacilli and bifidobacteria more frequently ($p < 0.001$) colonized the guts of supplemented infants. Probiotic treatment had no effect on the incidence of all allergic diseases by age two years but significantly prevented eczema and especially atopic eczema. Results suggested an inverse association between atopic diseases and colonization of the gut by probiotics.

Peluso et al., (2007) examined whether different strains of lactobacilli could directly regulate the functions of human T cells [136]. $CD4^+$ T cells were isolated from blood and intestinal lamina propria (LP) of normal individuals and patients with inflammatory bowel disease (IBD). Mononuclear cells were also isolated from Peyer's patches. Cells were activated with anti-CD3/CD2/CD28 in the presence or absence of *Lactobacillus paracasei* subsp. *paracasei* B21060, *Lactobacillus paracasei* subsp. *paracasei* F19, or *Lactobacillus casei* subsp. *casei* DG. Cell proliferation and death, Foxp3, intracellular pH, and cytokine production were evaluated by flow cytometry. *Lactobacillus paracasei* subsp. *paracasei* B21060 but neither *Lactobacillus paracasei* subsp. *paracasei* F19 nor *Lactobacillus casei* subsp. *casei* DG inhibited blood $CD4^+$ T-cell growth. This effect was associated with no change in cell survival, expression of Foxp3, or production of INF-γ, IL-4, IL-5, and IL-10. *Lactobacillus paracasei* subsp. *paracasei* B21060-mediated blockade of $CD4^+$ T-cell proliferation required a viable bacterium and was associated with decreased MCT-1 expression and low intracellular pH. *Lactobacillus paracasei* subsp. *paracasei* B21060 also inhibited the growth of Peyer's patch mononuclear cells, normal lymphocytes, and IBD $CD4^+$ LP lymphocytes without affecting cytokine production. The data showed that *Lactobacillus paracasei* subsp. *paracasei* B21060 blocks T-cell growth, thus suggesting a mechanism by which these probiotics could interfere with T-cell-driven immune responses.

Klein et al., (2008) determined whether a combination of *Lactobacillus acidophilus* 74-2 and *Bifidobacterium animalis* subsp *lactis* DGCC 420 (could affect the faecal microbiota as well as immunological parameters and blood lipids in healthy adults [137]. It was comducted a placebo-controlled, double-blinded, randomized crossover trial on 26 healthy volunteers (mean age 25 years) and all of them completed the study. After 3-week run-in period, half of the volunteers consumed 300 g/day of yoghurt supplement containing probiotic strains *Lactobacillus acidophilus* 74-2 and *Bifidobacterium animalis* subsp *lactis* 420, and the other half received the placebo product for a period of 5 weeks. The two groups were crossed during the following 5-week period. Blood and faecal samples were collected at the end of each period. The faecal content of probiotic bacteria, faecal short-chain fatty acids (SCFA), serum lipids and plasma immune system biomarkers were evaluated. Faecal proportions of *Lactobacillus acidophilus* and of *Bifidobacterium animalis* subsp *lactis* increased significantly from 0.02 to 0.19 and 0.4 to 1.4% ($p < 0.05$), respectively.

Percentages of granulocytes and monocytes showing phagocytic activity were significantly elevated from 92 to 95% during probiotic intervention, whereas their oxidative burst activity and specific immune parameters remained unaffected. Fecal SCFA and serum cholesterol levels were not influenced by the probiotics. However, serum concentrations of triacylglyceroles decreased significantly by 11.6% ($p < 0.05$) in the probiotic supplementation period. *Lactobacillus acidophilus* and *Bifidobacterium animalis* subsp *lactis* were recovered in faeces in significantly elevated numbers after supplementation. They were able to modulate unspecific cellular immune response indicated by the increased phagocytic activity.

Feighery et al., (2008) studied the probiotic, *Lactobacillus salivarius* subsp. *salivarius* 433118 (UCC118), for its potential to attenuate colitis, modulate immune responses and alter intestinal barrier dysfunction in two different mouse models of inflammatory bowel disease (IBD) [138]. Following oral treatment with UCC118, faecal microbial analysis indicated that viable intact bacteria reached the colons of interleukin (IL)-$10^{-/-}$ mice and dextran sodium sulphate (DSS)-treated mice. Neither prophylactic nor therapeutic UCC118 treatment significantly prevented or attenuated inflammation in either model. In all studies, the probiotic-treated mice had comparable cytokine responses as vehicle-treated animals. Mannitol permeability was increased across colonic mucosae mounted in Ussing chambers from DSS-treated mice, but not in IL-$10^{-/-}$ mice. However, colonic mucosae from UCC118-treated mice had unchanged transepithelial electrical resistance (TEER) values and mannitol fluxes compared to controls. In two different mouse colitis models examined under a range of histological and functional criteria, *Lactobacillus* spp. had limited potential as a prophylactic or therapeutic treatment for inflammatory bowel disease. While several studies showed therapeutic activity for this probiotic in mouse models of IBD, these data suggested that there were inter-study variables in formulation, study design, animal models and assessment criteria that could impact on interpretation of probiotic efficacy.

Imaoka et al., (2008) determined the anti-inflammatory activity of probiotic Bifidobacteria in Bifidobacteria-fermented milk (BFM) which was effective against active ulcerative colitis (UC) and exacerbations of UC, and explored the immunoregulatory mechanisms [139]. Peripheral blood mononuclear cells (PBMNC) from UC patients or HT-29 cells were co-cultured with heat-killed probiotic bacteria or culture supernatant of *Bifidobacterium breve* strain Yakult (BbrY) or *Bifidobacterium bifidum* strain Yakult (BbiY) to estimate the amount of IL-10 or IL-8 secreted (figure 52 A-B). Both strains of probiotic Bifidobacteria contained in the BFM induced IL-10 production in PBMNC from UC patients, though BbrY was more effective than BbiY.

Conditioned medium (CM) and DNA of both strains inhibited IL-8 secretion in HT-29 cells stimulated with TNF-α, whereas no such effect was observed with heat-killed bacteria. The inhibitory effect of CM derived from BbiY was greater than that of CM derived from BbrY.

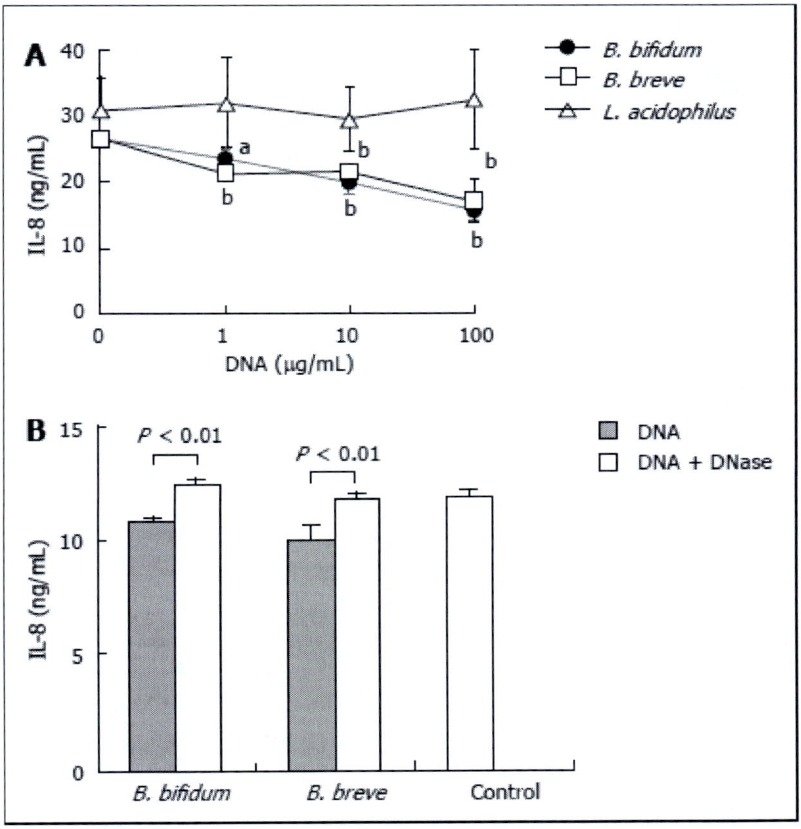

Figure 52. Effects of probiotic DNAs on IL-8 production in HT-29 cells. HT-29 cells were stimulated with TNF-α in the absence or presence of various concentrations of probiotic DNA without (A) or with DNase treatment (B). $^{a}p < 0.05$, $^{b}p < 0.01$ vs control [139].

DNAs of the two strains had a comparable inhibitory activity against the secretion of IL-8. CM of BbiY induced a repression of IL-8 gene expression with a higher expression of IκB-ς mRNA four hours after culture of HT-29 cells compared to that in the absence of CM. Probiotic *Bifidobacterium* strains in BFM enhance IL-10 production in PBMNC and inhibit IL-8 secretion in intestinal epithelial cells, suggesting that BFM had anti-inflammatory effects against ulcerative colitis.

West et al., (2008) aimed at determining the impact of *Lactobacillus* F19 (LF19) during weaning on infections and IgG antibody responses to routine vaccines [140]. In a double-blind, placebo-controlled randomized intervention trial, infants were fed cereals with (n = 89) or without LF19 (n = 90) from 4 to 13 months of age. Infants were immunized with DTaP (diphtheria and tetanus toxoid and acellular pertussis), polio and Hib-conjugate vaccines at (3), 5(1/2) and 12 months of age. The number of days with infections, antibiotic prescriptions and antibody concentrations to Hib capsular polysaccharide (HibPS), diphtheria toxin (D) and tetanus toxoid (T) before and after the second and third doses were assessed. Days with infectious symptoms did not differ between the groups. Days with antibiotic prescriptions were fewer in the LF19 group ($p = 0.044$). LF19 enhanced anti-D concentrations when adjusting for breastfeeding duration and colonization with LF19 ($p = 0.024$). There was an interaction of the intervention and colonization with LF19 on anti-T concentrations during the course of vaccination ($p = 0.035$). The anti-HibPS concentrations were higher after the first and second dose of Hib vaccine in infants breastfed < 6 months compared with those breastfed ≥ 6 months ($p < 0.05$), with no effect by LF19. Feeding LF19 did not prevent infections, but increased the capacity to raise immune responses to protein antigens, with more pronounced effects in infants breastfed < 6 months.

Vidal et al., (2008) examined the immunomodulatory capacity of the probiotic *Lactobacillus paracasei* NCC2461 combined or not with a prebiotic composition, FOS/inulin, in aged mice [141]. Male C57BL/6J mice (21-months-old) were allocated to one of three groups fed ad libitum for 44 days with different diets: a normal diet (control), a normal diet plus NCC2461 given in the drinking water, or a diet containing FOS/inulin plus NCC2461 in the drinking water. All mice were immunized on day 15 and challenged on day 22 with keyhole limpet hemocyanin (KLH). T helper T_H1 cell-dependent immune responses (anti-KLH immunoglobulin G(2a) [IgG(2a)] levels and delayed type hypersensitivity response) were increased significantly in NCC2461-supplemented mice when compared to controls. Supplementation with FOS/inulin did not further improve the immune-enhancing effect mediated by the probiotic.

Splenocyte proliferation, T cell subsets, systemic total IgG levels, and mucosal total IgA responses were not affected. Interestingly, supplementation with NCC2461 modulated the intestinal microbiota composition by increasing the numbers of Bifidobacteria and Lactobacilli. In conclusion, oral intake of *Lactobacillus paracasei* NCC2461 by aged mice enhanced the specific adaptive immune response to *in vivo* antigenic challenge without altering other cellular and humoral immune responses. The poor responsiveness to antigenic challenge, frequently observed in elderly people, could be improved by supplementation with *Lactobacillus paracasei* NCC2461.

Paturi et al., (2008) evaluated the immunostimulatory effects of *Lactobacillus acidophilus* LAFTI L10 and *Lactobacillus paracasei* LAFTI L26 to determine their probiotic properties for functional food applications [142]. Mice were given oral doses of either *Lactobacillus acidophilus* L10 or *Lactobacillus paracasei* L26 (10^8 CFU/50 μL/day), and the effects on immune responses and bacterial translocation were assessed after the 14-day feeding trial. The proliferative responses of splenocytes to concanavalin A and lipopolysaccharide were significantly higher in mice fed *Lactobacillus acidophilus*. Concanavalin A-induced splenocyte proliferative responses increased significantly in mice fed *Lactobacillus paracasei*. Interleukin 10 and interferon gamma production from the splenocytes stimulated with concanavalin A were enhanced in mice fed *Lactobacillus acidophilus* or *Lactobacillus paracasei*. The phagocytic activity of the peritoneal macrophages was significantly higher in mice fed either *Lactobacillus acidophilus* or *Lactobacillus paracasei* compared with control mice. In mice fed *Lactobacillus acidophilus* or *Lactobacillus paracasei*, the bacterial translocation of *Lactobacillus* spp. and total anaerobes to Peyer's patches and mesenteric lymph nodes was modulated compared with that in the control mice. Furthermore, there was no indication of disruption of intestinal mucosal integrity and thus no bacterial translocation to spleen, liver, or blood in mice fed either *Lactobacillus acidophilus* or *Lactobacillus paracasei*. The results of this study indicate that *Lactobacillus acidophilus* and *Lactobacillus paracasei* were potential enhancers of systemic immunity and were nonpathogenic, as suggested by their bacterial translocation profiles in healthy mice.

Tsai et al., (2008) investigated the possible time-dependent role of LAB in immunomodulation in BALB/c mice fed daily with *Lactobacillus paracasei* subsp. *paracasei* NTU 101 (10^8 CFU) for 3, 6, and 9 weeks, and following feeding with *Lactobacillus*-free food for a further 7 days [143]. Upregulation of the antigen-presenting ability of dendritic cells and expression of natural killer group-2 D (NKG_2D) molecules capable of trigger natural killer cell-mediated cytotoxicity was observed. Lymphocyte proliferation and antibody production were also significantly increased in mice after treatment (figure 53). Innate and adaptive immunity remained constant even at the most protracted feeding time, indicative of the time dependence of the bacterial-mediated enhanced immunity. To better correlate intestinal microflora with immunity, the intestinal contents of probiotics and harmful microorganisms were determined. Results showed an altered intestinal microflora, with increases in bifidobacteria and lactobacilli and a decreased content of *Clostridium perfringens* after feeding with *Lactobacillus paracasei* subsp.

paracasei NTU 101. It could be possible that persistent activation of immunity might be induced by intestinal probiotics.

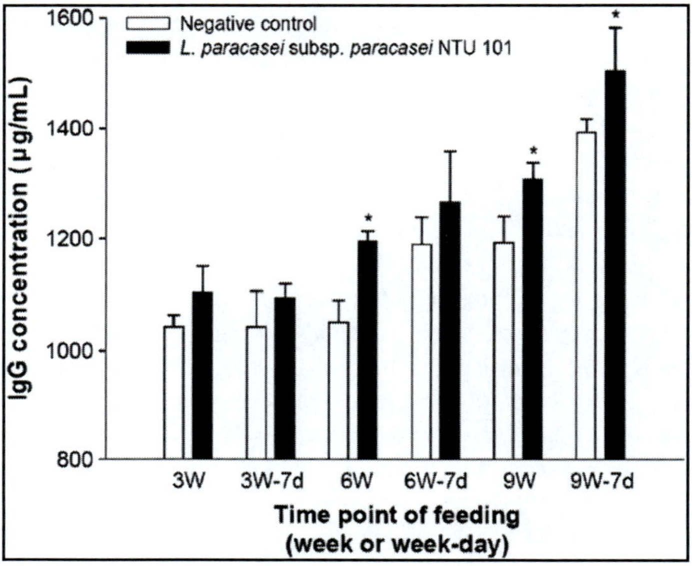

Figure 53. Production of IgG antibodies at different feeding times from BALB/c mice after orally administered with *Lactobacillus paracasei* subsp. *paracasei* NTU 101. All data are presented as mean ± SD (n = 6). The experiment group is depicted in black and the control group is depicted in white. * represents significantly different from negative control ($p < 0.05$). 3W, 6Wand 9W represent the antibody concentration change after oral administration for 3, 6 and 9 wks. 3W-7d, 6W-7d and 9W-7d represent the antibody concentration change after oral administration for 3, 6 and 9 wks and then termination for 7 days [143].

Talebi et al., (2008) investigated the effects of the probiotic PrimaLac (*Lactobacillus acidophilus*, *Lactobacillus casei*, *Enterococcus faecium* and *Bifidobacterium bifidium*) on performance and immune responses to vaccination against Newcastle disease and infectious bursal disease in broiler chickens [144]. Use of PrimaLac significantly ($p < 0.05$) enhanced broiler performance by improving body weight and decreasing the feed conversion ratio. Although administration of the probiotic appeared to improve the antibody responses to Newcastle disease virus and infectious bursal disease vaccination, the antibody titres of the probiotic-treated group were not significantly different from those not receiving probiotics.

Reyes-Becerril et al., (2008) evaluated the effects of dietary administration of the live yeast *Debaryomyces hansenii* CBS 8339 on the gilthead seabream (*Sparus aurata* L.) innate immune responses [145]. Seabream were fed control or *Debaryomyces hansenii*-supplemented diets (10^6 CFU/g) for 4 weeks. Humoral (seric alternative complement and peroxidase activities), and cellular (peroxidase, phagocytic, respiratory burst and cytotoxic activities) innate immune parameters and antioxidant enzymes (superoxide dismutase (SOD) and catalase (CAT) were measured from serum, head-kidney leucocytes and liver, respectively, after 2 and 4 weeks of feeding (figure 54 A). Expression levels of immune-associated genes, Hep, IgM, TCR-β, NCCRP-1, MHC-IIα, CSF-1R, C3, TNF-α and IL-1β, were also evaluated by real-time PCR in head-kidney, liver and intestine. Humoral immune parameters were not significantly affected by the dietary supplementation of yeast at any time of the experiment. On the other hand, *Debaryomyces hansenii* administration significantly enhanced leucocyte peroxidase and respiratory burst activity at week 4. Phagocytic and cytotoxic activities significantly increased by week 2 of feeding yeast but unchanged by week 4 (figure 54 B-C). A significant increase in liver SOD activity was observed at week 2 of feeding with the supplemented diet; however CAT activity was not affected by the dietary yeast supplement at any time of the experiment. Finally, the yeast supplemented diet down-regulated the expression of most seabream genes, except C3, in liver and intestine and up-regulated all of them in the head-kidney (figure 54 D). Results strongly supported the idea that live yeast *Debaryomyces hansenii* strain CBS 8339 could stimulate the innate immune parameters in seabream, especially at cellular level.

Ghadimi et al., (2008) sought to assess the effect of different probiotics, *Lactobacillus rhamnosus* GG, *Lactobacillus gasseri* (PA16/8), *Bifidobacterium bifidum* (MP20/5), and *Bifidobacterium longum* (SP07/3), on the T_H1 and T_H2 responses of peripheral blood mononuclear cells (PBMCs) from healthy subjects and from patients with allergy against house dust mite to *Staphylococcus* enterotoxin A (SEA) and *Dermatophagoides pteronyssinus* (Dpt) [146].

To elucidate the molecular basis of these effects, the effects of bacterial genomic DNA were compared with the effects of viable bacteria. PBMCs from allergic patients and from healthy donors were incubated for 24 or 48 hours, respectively, with or without SEA and Dpt allergens. The effects of preincubation with live probiotic bacteria and the effect of their genomic DNA, added simultaneously to cultures and incubated for 24 h, were assessed by measuring T_H1/T_H2-cytokine production.

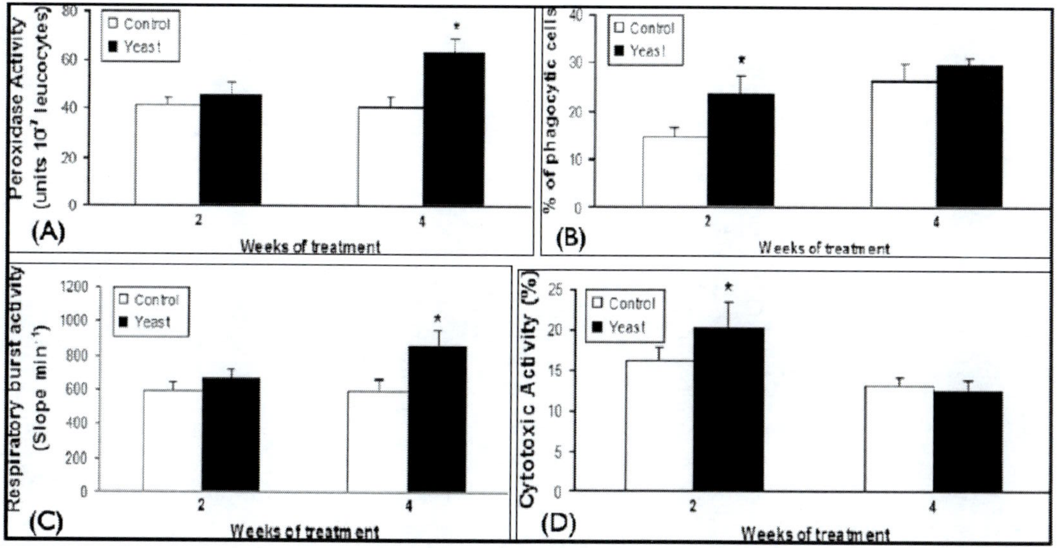

Figure 54. (A) Leucocyte peroxidase activity from gilthead seabream fed non-supplemented (control) and *Debaryomyces hansenii* -supplemented diets (10^6 CFU/g). Data represent the mean ± SD (n = 10). Asterisk denotes statistically significant differences between control and yeast-fed groups ($p \leq 0.05$); (B) phagocytosis of FITC-labelled *Saccharomyces cerevisiae* by head-kidney leucocytes from gilthead seabream fed nonsupplemented (control) and *Debaryomyces hansenii*-supplemented diets (10^6 CFU/g). Data represent the mean ± SD (n = 10). Asterisk denotes statistically significant differences between control and yeast-fed groups ($p \leq 0.05$); (C) respiratory burst activity of head-kidney leucocytes from gilthead seabream fed non-supplemented (control) and *Debaryomyces hansenii*-supplemented diets (10^6 CFU/g). Data represent the mean ± SD (n = 10). Asterisk denotes statistically significant differences between control and yeastfed groups ($p \leq 0.05$); (D) natural cytotoxic (NCC) activity of head-kidney leucocytes from gilthead seabream specimens fed non-supplemented (control) and *Debaryomyces hansenii*-supplemented diets (10^6 CFU/g). Data represent the mean ± SD (n = 10). Asterisk denotes statistically significant differences between control and yeast-fed groups ($p \leq 0.05$) [145].

The tested live Gram-positive probiotic bacteria and their genomic DNA inhibited SEA- and Dpt-stimulated secretion of T_H2-cytokines (IL-4 and IL-5) and enhanced the stimulation of IFN-γ. This effect was dose-dependent with a dosage-optimum, which was identical for all lactic acid producing bacteria (LAB) tested (10 bacteria per PBMC) and their DNA (75 ng/mL).

Based on the maximal effects achieved with LAB and their DNA, more than 50% of the effects seem to be contributed by DNA. No significant effect was induced by the control, Gram-negative *Escherichia coli* TG1. Lactobacilli and Bifidobacteria reduced SEA-stimulated IL-4 and IL-5 production more effectively in PBMCs from healthy subjects than from allergic patients. In contrast to this, inhibition of Dpt-stimulated IL-4- and IL-5-secretion was more pronounced in cells from allergic subjects. Compared with living LAB, DNA inhibited IL-4- and IL-5-secretion in a similar manner. SEA- and even more so Dpt-stimulated IFN-γ stimulation by living LAB was less pronounced in allergic than in healthy subjects, whereas IFN-γ stimulation by their DNA was more pronounced in allergic subjects (figure 55).

The tested probiotic bacteria as well as their genomic DNA modulated the T_H1/T_H2 response to some allergens dose-dependently. DNA seems to contribute to 50% of the effect exerted by living bacteria in this *in vitro* model. The magnitude of the probiotic effects differed between healthy and allergic subjects. Candela et al., (2008) investigated the protective role toward enteropathogen infection of probiotic strains belonging to *Lactobacillus* and *Bifidobacterium* [147]. According to experimental data, *Lactobacillus acidophilus* Bar13, *Lactobacillus plantarum* Bar10, *Bifidobacterium longum* Bar33 and *Bifidobacterium lactis* Bar30 were effective in displacing the enteropathogens *Salmonella typhimurium* and *Escherichia coli* H10407 from a CaCO-2 cell layer (figure 56). Moreover, *Lactobacillus acidophilus* Bar13 and *Bifidobacterium longum* Bar33 have been assessed for their immunomodulatory activity on IL-8 production by HT-29 cells. Both strains showed the potential to protect enterocytes from an acute inflammatory response. These probiotic strains are potential candidates for the development of new functional foods helpful in counteracting enteropathogen infections.

Takagi et al., (2008) compared differences among six representative strains of lactobacilli in regard to their ability to stimulate bone marrow cell-derived dendritic cells (BMDCs) *in vitro* and tumor suppression *in vivo* [148].

BM-DCs were co-cultured with a *Lactobacillus* strain *in vitro*, and the IL-12 released into the culture supernatant was measured by enzyme-linked immunosorbent assay. Tumors were chemically induced by a single subcutaneous injection of 3-methylcholanthrene (MC) in BALB/c mice.

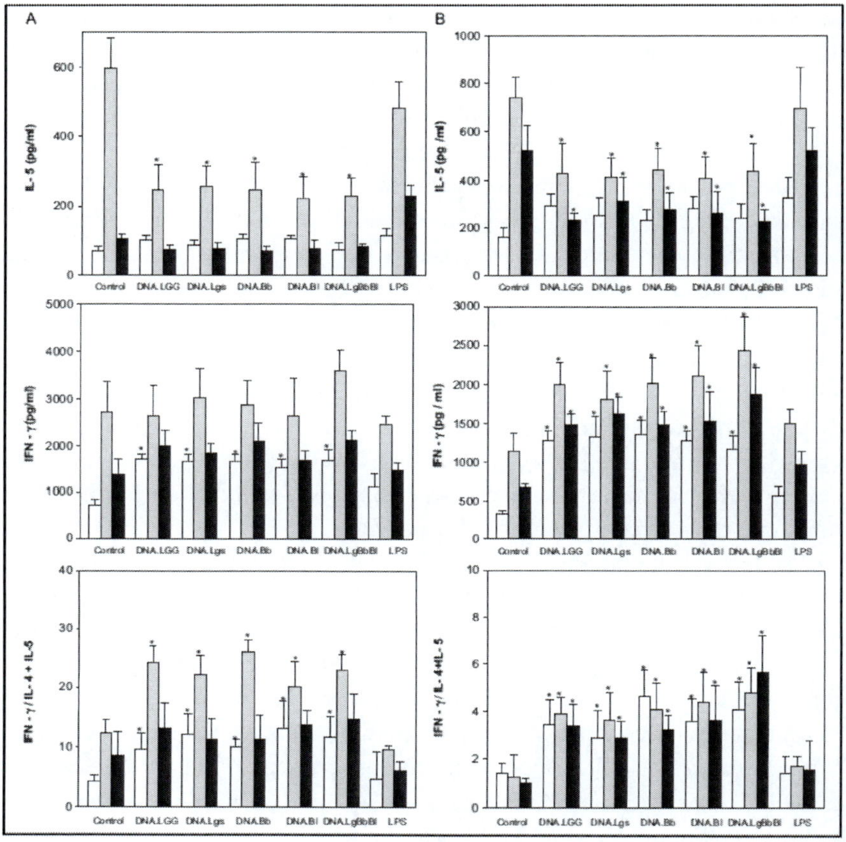

Figure 55. Inhibitory effect of genomic DNA from probiotic bacteria on IL-4, IL-5, and IFN-γ production by SEA or Dpt-stimulated PBMC from healthy (A, n = 5) and allergic (B, n = 8) subjects. PBMC (2 x 10^6 cells/mL) were incubated for 24 h with or without (control) genomic DNA from four strains of probiotic bacteria, *Lactobacillus rhamnosus* GG, *Lactobacillus gasseri* (PA16/8), *Bifidobacterium bifidum* (MP20/5), and *Bifidobacterium longum* (SP07/3) and LgBbBl (a mix of *Lactobacillus gasseri* (PA16/8), *Bifidobacterium bifidum* (MP20/5), and *Bifidobacterium longum* (SP07/3)) at a concentration of 75 μg/mL before stimulation with SEA (2 μg/mL) or Dpt (2 μg/mL). LPS was used at a concentration of 100 μg/mL. IL-4, IL-5, and IFN-γ production were quantified after 24 h incubation by means of specific ELISA. Asterisks indicate a significant inhibition (*$p < 0.05$, **$p < 0.01$) of cytokine production in comparison with control [146].

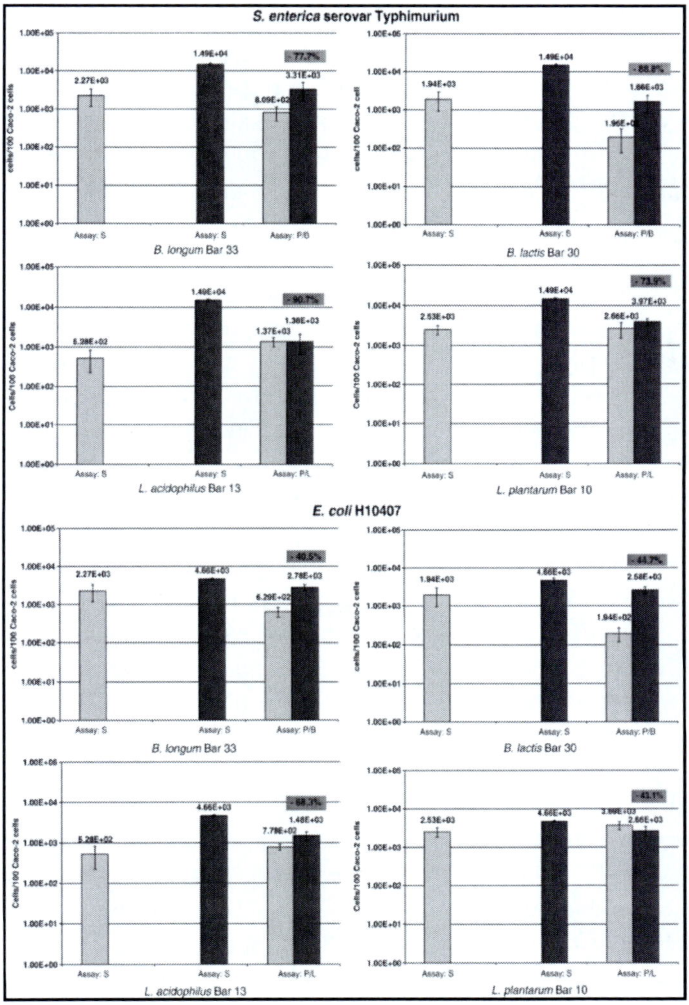

Figure 56. Displacement activity of the bifidobacteria (B) (*Bifidobacterium longum* Bar33 and *Bifidobacterium lactis* Bar30), and *Lactobacillus* (L) (Continued)

(*Lactobacillus acidophilus* Bar13 and *Lactobacillus plantarum* Bar10) strains toward *Salmonella typhimurium* and *Escherichia coli* H10407 (P). For each couple pathogen probiotic bacteria, the Log values of bacteria bound to 100 CaCO-2 cells, as obtained in assays containing the single strains (Assays S) and the displacement assays (Assay P/B or P/L), are given. The error bars represent ± SD of the mean values. Grey columns represent the pathogen, white columns the respective *Bifidobacterium* or *Lactobacillus* spp [147].

The test diets containing *Lactobacillus* strains were given from the day of the MC injection, and the tumor incidences were monitored. Peyer's patches were dissected from *Lactobacillus*-fed mice, and the status of c-Src, a regulator of DCs, in Peyer's patch cells was examined by Western blotting. In the co-culture system, *Lactobacillus fermentum* FERM P-13857 and LcS potently elicited IL-12 production. LcS but not the other strains of lactobacilli showed tumor suppression. The inactive form of c-Src, phosphorylated at Tyr 527, was dominantly detected in Peyer's patches resected from *Lactobacillus fermentum* FERM P-13857-fed mice compared with LcS-fed mice. The responses of DCs could be associated with tumor suppression by an ingested *Lactobacillus* strain.

Kekkonen et al., (2008) evaluated the effects of three potentially anti-inflammatory probiotic bacteria from three different genera on immune variables in healthy adults in a clinical setting based on previous *in vitro* characterization of cytokine responses [149]. A total of 62 volunteers were enrolled in a randomized, double-blind and placebo-controlled parallel group intervention study.

The volunteers were randomized to receive a milk-based drink containing either *Lactobacillus rhamnosus* GG (LGG), *Bifidobacterium animalis* ssp. lactis Bb12 (Bb12), or *Propionibacterium freudenreichii* subsp. *shermanii* JS (PJS) or a placebo drink for 3 weeks. Venous blood and saliva samples were taken at baseline and on d 1, 7 and 21 (figure 57). Fecal samples were collected at baseline and at the end of intervention. The serum hsCRP expressed as the median AUC0-21 (minus baseline) was 0.018 mg/L in the placebo group, - 0.240 mg/L in the LGG group, 0.090 mg/L in the Bb12 group and -0.085 mg/L in the PJS group ($p = 0.014$). *In vitro* production of TNF-α from *in vitro* cultured peripheral blood mononuclear cells (PBMC) was significantly lower in subjects receiving LGG vs placebo.

IL-2 production from PBMC in the Bb12 group was significantly lower compared with the other groups. In conclusion, probiotic bacteria have strain-specific anti-inflammatory effects in healthy adults.

Paineau et al., (2008) investigated the immunomodulatory properties of seven probiotic strains (*Bifidobacterium lactis* Bi-07, *Bifidobacterium lactis* Bl-04, *Lactobacillus acidophilus* La-14, *Lactobacillus acidophilus* NCFM, *Lactobacillus plantarum* Lp-115, *Lactobacillus paracasei* Lpc-37, *Lactobacillus salivarius* Ls-33) [150]. Eighty-three healthy volunteers aged 18–62 years consumed 2×10^{10} CFU of bacteria or a placebo (maltodextrin)

	Placebo (*n* = 16)	LGG (*n* = 13)	Bb12 (*n* = 16)	PJ5 (*n* = 17)	*P* value[1]	Localization
TNF-α						
Streptococcus	703 (-315-1784)	-1883 (-2389-540)	-645 (-1843-1403)	315 (-1045-1460)	0.025	LGG *vs* placebo
Influenza	31 (-3-66)	6 (-83-84)	-27 (-101-37)	29 (-39-83)	0.32	
LPS	11 (-16-38)	-15 (-31-10)	14 (-36-48)	-6 (-78-36)	0.53	
IFN-γ						
Streptococcus	-19 (-221-97)	-10 (-227-284)	71 (-161-360)	-23 (-223-155)	0.6	
Influenza	117 (29-284)	-72 (-221-194)	-7 (-436-189)	102 (-59-218)	0.25	
LPS	NA	NA	NA	NA		
IL-1β						
Streptococcus	2308 (45-5222)	-1324 (-5609-352)	444 (-6152-5000)	649 (-3412-3747)	0.49	
Influenza	166 (13-478)	83 (-183-273)	-15 (-817-170)	156 (-75-791)	0.69	
LPS	67 (-29-138)	5 (-147-114)	51 (-40-122)	17 (-43-199)	0.66	
IL-2						
Streptococcus	5 (-23-79)	-46 (-176-0)	1 (-135-97)	39 (-105-213)	0.33	
Influenza	25 (-18-42)	14 (-51-62)	-55 (-162-14)	10 (-27-39)	< 0.001	Bb12 *vs* others
LPS	2 (-18-35)	27 (-25-57)	1 (-42-51)	6 (-35-25)	0.54	
IL-6						
Streptococcus	793 (-691-4829)	-379 (-3955-397)	205 (-2757-1553)	175 (-4344-909)	0.82	
Influenza	1530 (166-4632)	1288 (-4329-4073)	550 (-7178-1905)	1644 (-327-3344)	0.74	
LPS	947 (-418-2527)	-2189 (-3675-4053)	329 (-1949-3950)	639 (-1217-1607)	0.13	
IL-8						
Streptococcus	240 (-1585-3914)	34 (-3066-3143)	400 (-3638-2891)	132 (-2667-2976)	0.96	
Influenza	-455 (-1966-1670)	-193 (-2953-1520)	-1675 (-4245 to -601)	-1148 (-2550-1707)	0.31	
LPS	-742 (-3384-180)	-149 (-1608-1517)	-334 (-2692-739)	-1111 (-2457-836)	0.78	
IL-10						
Streptococcus	907 (263-2149)	-4 (-881-2420)	452 (-2982-1735)	226 (-86-950)	0.29	
Influenza	95 (48-284)	-57 (-159-233)	4 (-301-130)	95 (13-350)	0.25	
LPS	381 (19-602)	78 (-298-656)	347 (-403-796)	187 (28-1440)	0.51	
IL-12						
Streptococcus	26 (-18-75)	-32 (-99-36)	22 (-36-67)	35 (-73-172)	0.46	
Influenza	0 (-4-16)	7 (-9-30)	8 (-19-51)	5 (-25-37)	0.88	
LPS	15 (1-36)	3 (-7-28)	0 (-6-29)	0 (-62-20)	0.23	

Figure 57. The effect of a 3-wk probiotic intervention on *in vitro* cytokine production (pg/mL) in peripheral blood mononuclear cells stimulated with *Streptococcus pyogenes*, LPS from *Escherichia coli* and Influenza A H3N2 virus of healthy adults (n = 62) presented as median AUC0-21 minus baseline (IQR).

	Absolute changes in immunoglobulin between D0 and D21				Absolute changes in immunoglobulin between D21 and D28			
	Serum IgA (OD_{450nm})	Serum IgG (OD_{450nm})	Serum IgM (OD_{450nm})	Salivary IgA (OD_{450nm})	Serum IgA (OD_{450nm})	Serum IgG (OD_{450nm})	Serum IgM (OD_{450nm})	Salivary IgA (OD_{450nm})
Bifidobacterium lactis	−13.8	8.9[d]	5.7	0.08	−0.3	2.8	−5.8	0.06
Bi-07	(−71.3/7.3)	(−666.0/125.3)	(−14.6/54.6)	(−0.3/0.2)	(−139.67/147.3)	(−300.6/442.3)	(−60.6/102.0)	(−0.3/0.7)
Bifidobacterium lactis	6.0	29.0[a,b]	2.3	0.06	0.3	−4.3	−2.6	−0.01
Bl-04	(−108.0/358.6)	(−15.1/1771.0)	(−31.0/53.3)	(−0.2/0.4)	(−140.3/445.1)	(−149.0/29.0)	(−135.0/29.6)	(−0.3/0.4)
Lactobacillus acidophilus	−30.0[d,e]	32.3[a,e]	−19.6[e]	0.00	−11.0[d]	8.6	3.6	0.08[c]
La-14	(−176.3/76.0)	(−84.3/450.0)	(−261.3/18.5)	(−0.1/0.2)	(−84.3/64.6)	(−355.0/477.3)	(−23.3/125.6)	(−0.0/0.4)
Lactobacillus acidophilus	−46.5	−1.0	−2.8	0.00	50.6[c,d]	14.3	15.8[c]	0.02
NCFM®	(−309.3/390.3)	(−426.0/164.6)	(−41.6/94.6)	(−0.0/0.3)	(−6.0/440.0)	(−16.1/728.3)	(−5.5/465.0)	(−0.3/0.0)
Lactobacillus plantarum	24.6	40.6[d,e]	4.3	−0.02	−16.6	−1.3	−3.6	0.03
Lp-115	(−206.0/616.6)	(−19.3/616.6)	(−47.3/61.0)	(−0.1/0.2)	(−241.6/59.3)	(−141.3/76.0)	(−146.6/27.3)	(−0.1/0.6)
Lactobacillus paracasei	−3.3	−2.3	9.6	0.02	−17.6	24.6	−11.0[d,f]	0.01
Lpc-37	(−125.3/235.3)	(−117.6/61.1)	(−47.3/73.3)	(−0.0/0.6)	(−167.6/392.0)	(−26.6/439.8)	(−61.6/443.3)	(−0.0/0.0)
Lactobacillus salivarius	31.0	−10.3	17.6	−0.04[a]	0.0	−7.6	15.6	0.00
Ls-33	(−56.3/953.0)	(−284.6/203.6)	(−82.6/376.3)	(−0.2/0.0)	(−902.0/310.6)	(−304.0/73.5)	(−386.3/178.6)	(−0.0/0.3)
Placebo	0.6	−14.0	11.6	0.06[b]	3.1	33.6	5.6	0.05[c]
	(−136.0/25.0)	(−594.6/383.3)	(−60.3/60.8)	(−0.3/0.3)	(−64.8/599.3)	(−225.6/1210.0)	(−64.5/166.0)	(−0.5/0.6)

Figure 58. Ig evolutions during the periods D0–D21 and D21–D28 [150].

over 3 weeks (D0–D21). Subjects received an oral cholera vaccine at D7 and at D14; blood and saliva samples were collected at D0, D21 and D28. Serum samples were analyzed for specific IgA, IgG and IgM, and saliva samples were analyzed for specific IgA only, by ELISA. Statistical analyses were based on Wilcoxon's signed-rank test (intragroup analyses) and exact median t-test (intergroup analyses). Salivary analysis showed no difference in specific IgA concentrations between groups. Serum analysis indicated an effect of some of the tested strains on specific humoral responses. Between D0 and D21, IgG increased in two probiotic groups, for example, *Bifidobacterium lactis* Bl-04 and *Lactobacillus acidophilus* La-14, compared with controls ($p = 0.01$) (figure 58). Trends toward significant changes in immunoglobulin serum concentrations compared with controls ($p < 0.1$) were found for six out of the seven probiotic strains. Some strains of probiotics demonstrated a faster immune response measured with serum immunoglobulin indicators, especially IgG, although overall vaccination was not influenced. Specific strains of probiotics could thus act as adjuvants to the humoral immune response following oral vaccination.

Ya et al., (2008) isolated a novel LAB, *Lactobacillus casei* Zhang (LcZhang) from koumiss. LcZhang exhibited favorable probiotic properties, such as acid resistance, bile resistance, gastrointestinal (GI) colonization ability, etc. In order to examine the immunomodulatory qualities of LcZhang, it was administered to healthy mice with varying doses of either live or heat-killed and measured various parameters of the host immune response [151]. The study was performed in four separate experiments via oral administration of live and heat-killed LcZhang to BALB/c mice for several consecutive days. The immunomodulating capacity of LcZhang was investigated *in vivo* by analyzing the profile of cytokines (figure 59) T cell subpopulations, and immunoglobulin concentrations induced in blood serum and intestinal fluid in BALB/c mice. Only live bacteria elicited a wide range of immune responses, including the increased production of IFN-γ, and depression of TNF-α levels. In addition, IL-2 and IL-2 receptor gene transcription increased significantly, but the proportion of T cell subsets appeared to be unaffected.

LcZhang was capable of inducing gut mucosal responses by enhancing the production of secretory Immunoglobulin A (sIgA) as well influencing the systemic immunity via the cytokines released to the circulating blood. Hence the dose-dependent administration of LcZhang was capable of influencing immune responses, implying that it could be a valuable strain for probiotic use in humans.

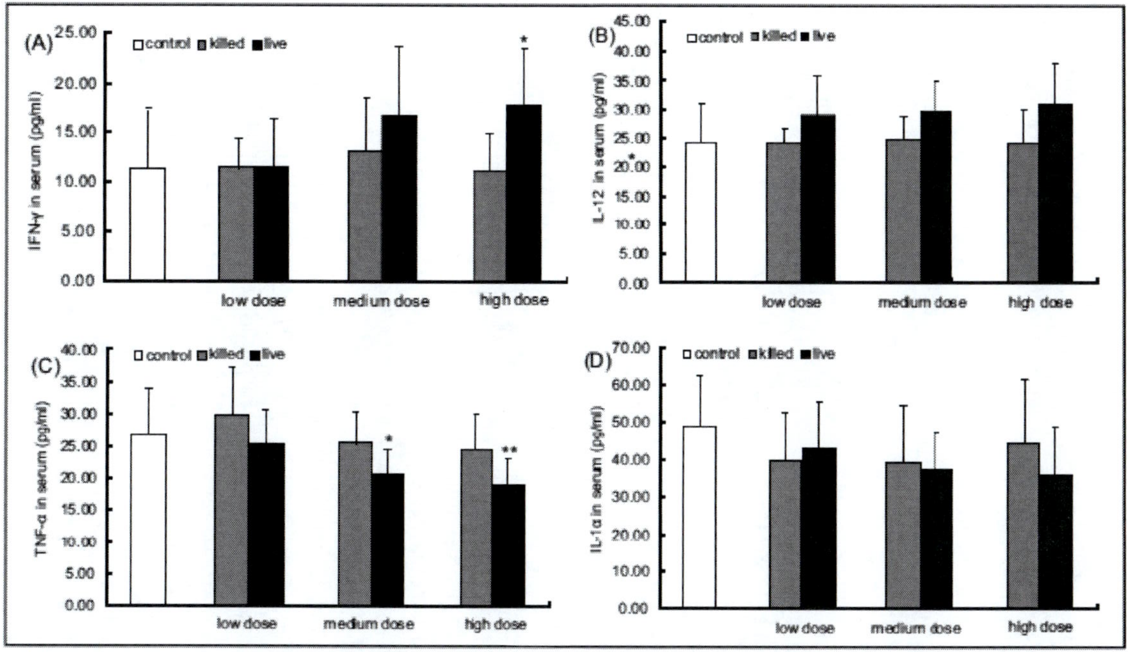

Figure 59. Effects of the oral administration of varying doses of LcZhang on cytokines secretion. Induction of major immunoregulatory cytokines (A: IFN-γ; B: IL-12; C: TNF-α; D: IL-1α) levels in serum by live LcZhang and heat-killed bacteria after 15 days oral administration. Data represent the mean ± standard deviation of each cytokine in pg/mL in serum for each test treatment (n = 10). Three dosages of bacteria were shown: low dose means 0.25×10^9 cells/mL; medium dose means 0.5×10^9 cells/ mL; high dose means 1.0×10^9 cells/mL. "Control" refers to the same volume of PBS (open column); "killed" represents heat killed LcZhang (grey column); "live" represents live LcZhang (solid column). Data are means ± SD (Error bars). $*p < 0.05$; $**p < 0.01$ [151].

Maassen et al., (2008) presented new data showing strain-specific differences on the effect of commercially available probiotic drinks (Danone: *Lactobacillus casei* DN 114-001, Vifit: *Lactobacillus rhamnosus* GG, Yakult: *Lactobacillus casei* Shirota) in an EAE rat autoimmune model for multiple sclerosis. In this particular model, these drinks did not enhance but rather suppress the disease [152].

Nandakumar et al., (2009) examined the effect of *Vibrio cholerae* and *Lactobacillus rhamnosus* GG on inflammatory chemokine gene expression in the HT-29 human intestinal epithelial cell line [153]. HT-29 cells were exposed to *Vibrio cholerae* 0139, *Lactobacillus rhamnosus* GG or both for 2 hours and cultured further thereafter for 4 hours. RNA was extracted from the cells and expression of genes for chemokines and related molecules was quantitated by real time PCR using a pathway focused PCR array (figure 60). TLR4 was silenced using shRNA and output of IL-8 into the media quantitated with and without *Vibrio cholerae* exposure. NFkB and p38 MAP kinase activation were determined by immunoblotting for IkBa and phosphorylated p38. *Vibrio cholerae* significantly upregulated gene expression for the neutrophil chemoattractant CXCL chemokines, IL-8, CXCL and CXCL in HT-29 cells, while downregulating the expression of macrophage- attracting C-C chemokines. TLR4 silencing did not reduce IL-8 output from HT-29 cells in response to *Vibrio cholerae*. IkB-α degradation (indicative of NFkB activation) was noted in the HT-29 cells soon after exposure to *Vibrio cholerae* and this recovered over time after removal of bacteria. p38 MAP kinase activation was not noted. *Vibrio cholerae* upregulated the expression of neutrophil attractant chemokines, most prominently IL-8, in HT-29 cells, but downregulated macrophage-attracting chemokines. Probiotic lactobacilli modulated the IL-8, but not the other chemokine gene changes, in response to *Vibrio cholerae*.

In 2009, Mileti et al. analyzed the immunological activity of three strains of lactobacilli (*Lactobacillus plantarum* NCIMB8826, *Lactobacillus rhamnosus* GG (LGG), *Lactobacillus paracasei* B21060) plus pathogenic *Salmonella typhimurium* (SL1344) to predict their *in vivo* efficacy in protecting against experimental colitis [154].

They studied the stimulatory effects of these different strains upon dendritic cells (DCs) either directly by co-culture or indirectly via conditioning of an epithelial intermediary (figure 61). Furthermore, they characterized the effects of these strains *in vivo* using a Dextran Sulphate Sodium (DSS) model of colitis. The three strains exhibited different abilities to induce inflammatory cytokine production by DCs with *Lactobacillus* plantarum being the most effective followed by LGG and *Lactobacillus paracasei.*

Gene		Fold up- or down-regulation		P-value	
		Lactobacillus	*Vibrio cholerae*	*Lactobacillus*	*Vibrio cholerae*
CXC ligands					
CXCL8	IL-8	1.87	42.3	0.3802	0.015
CXCL10	IP-10	3.87	16.07	0.766	0.013
CXCL1	GRO-α	1.08	14.55	0.2859	0.03
CXCL5	ENA-78	1.16	−4.77	0.2439	0.0209
CC ligands					
CCL1	I-309	−1	−2.96	0.201	0.0122
CCL2	MCP-1	1.52	−2.98	0.2227	0.0096
CCL8	MCP-2	1.88	−2.71	0.4433	0.0514
CCL13	MCP-4	1.67	−3.18	0.3545	0.0238
CCL3	MIP-1 α	1.67	−3.13	0.326	0.0151
CCL4	MIP-1β	1.08	−3.87	0.2745	0.0299
CC receptor					
CCR8		−1.02	−3.73	0.2494	0.0358
CCR3		−1.2	−2.73	0.205	0.0584
Related genes					
TNF		3.08	6.17	0.6047	0.011
SLIT2		2.09	−2.8	0.4494	0.0653
TLR2		1.23	−3.2	0.3204	0.0739
TREM1		1.66	−3.33	0.3644	0.0388
TCP10		1.52	−3.39	0.3061	0.0292
MMP-7		1.97	4.65	0.3225	0.01

Figure 60. Changes (upregulation or downregulation) in expression of important chemokine gene groups brought about in HT-29 cells by infection with *Vibrio cholerae*. GRO, growth related oncogene; MCP, monocyte chemoattractant protein; MIP, macrophage inflammatory protein; ENA, epithelial neutrophil activated peptide. *p* values were calculated using paired t-tests [153].

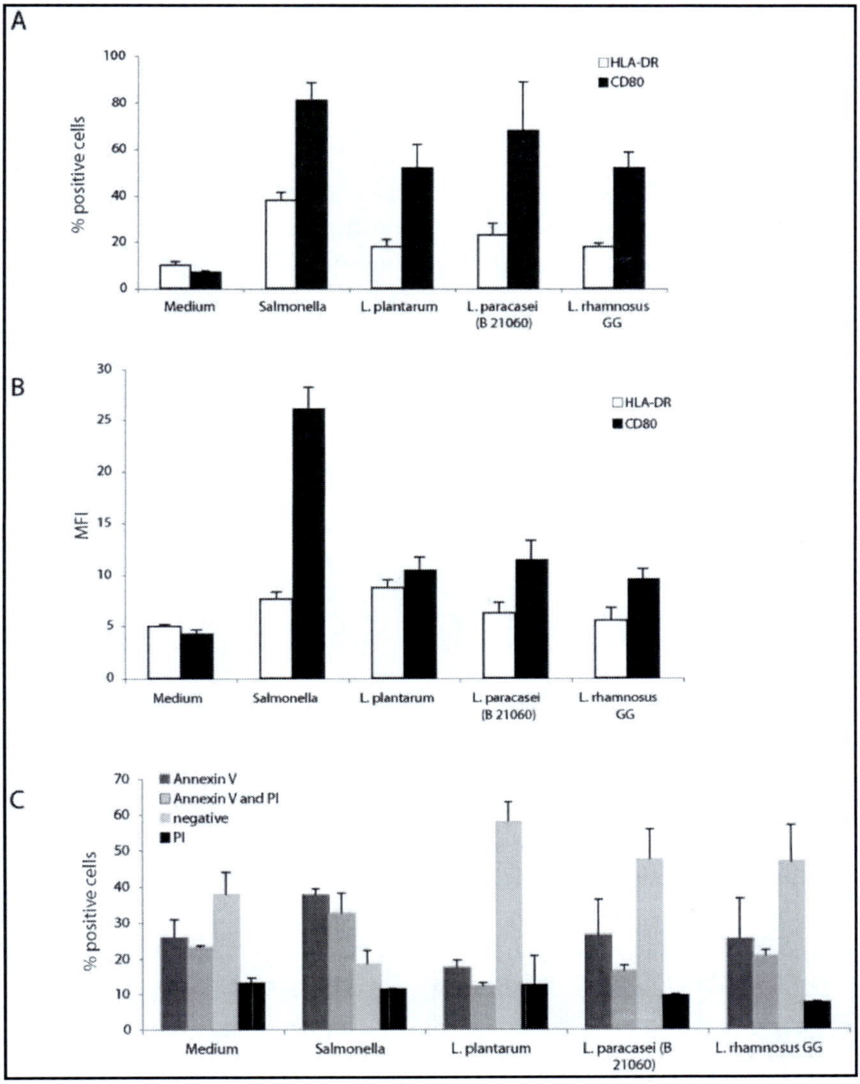

Figure 61. DCs are phenotypically similarly activated by lactobacilli. DCs were incubated or not with the reported live bacterial strains for 1 h in medium without antibiotics, washed and incubated for 23 h in medium with antibiotics. Cells were stained for HLA-DR and CD80 expression and analyzed by FACS. A. % of cells highly positive for the marker is reported. B. Mean fluorescence intensity (MFI) expression of markers is reported. C. Viability of the cells after 24 h incubation with bacteria. Cells were stained with propidium iodide (PI) and annexin V and analyzed by FACS [154].

Lactobacillus paracasei minimally induced the release of cytokines, while it also inhibited the potential of DCs to both produce inflammatory cytokines (IL-12 and TNF-α) and to drive T_H1 T cells in response to *Salmonella*. This effect on DCs was found under both direct and indirect stimulatory conditions and was dependent upon an as yet unidentified soluble mediator. When tested *in vivo*, *Lactobacillus plantarum* and LGG exacerbated the development of DSS-induced colitis and caused the death of treated mice, while, conversely *Lactobacillus paracasei* was protective. A new property of probiotics to either directly or indirectly inhibit DC activation by inflammatory bacteria was described. Moreover, some immunostimulatory probiotics not only failed to protect against colitis but amplified the disease progression.

Hemmerling et al., (2009) in a phase I dose-ranging safety trial tested the safety, tolerability, and acceptability of *Lactobacillus crispatus* CTV-05 (Lactin-V) administered by vaginal applicator [155]. Twelve healthy volunteers were enrolled in 3 blocks of 4 (5 x 10, 1 x 10, and 2 x 10 CFU/dose). Each block was randomized in a 3:1 ratio of active product to placebo. Participants used study product for 5 consecutive days, returned for follow-up on days 7 and 14, and had phone interviews on days 2 and 35. All twelve participants took 5 doses and completed study follow-up. Overall, 45 adverse events (AEs) occurred, of which 31 (69%) were genitourinary (GU) AEs. GU AEs appeared evenly distributed between the 3 treatment blocks and between Lactin-V and placebo arms. The most common GU AEs were vaginal discharge in 5 subjects (42%), abdominal pain in 4 subjects (33%), metrorrhagia in 4 subjects (33%), vulvovaginitis in 4 subjects (33%), vaginal candidiasis in 3 subjects (25%), and vaginal odor in 3 subjects (25%). Forty-one (91%) AEs were mild (grade 1) in severity. All 4 moderate AEs (grade 2) were unrelated to product use. No grade 3 or 4 AEs or serious adverse events (SAE) occurred. Laboratory parameters and colposcopy findings were within normal limits or clinically insignificant. The product was well tolerated and accepted. All three dose levels of Lactin-V appeared to be safe and acceptable in healthy volunteers.

Yang et al., (2009) addressed whether elevated *Bifidobacterium* titers relate positively to immune function [156]. The immunoactivities of 2 *Bifidobacterium* strains (*Bifidobacterium adolescentis* BBMN23 and *Bifidobacterium longum* BBMN68) isolated from healthy centenarians in China were evaluated. Different dosages (2×10^{11}, 2×10^9, or 2×10^7 CFU/kg body weight) of live Bifidobacteria were orally administered once per day to healthy BALB/c mice, and the control group was given sterile skim milk every day. After 4 weeks, the immune parameters including cellular immunity (delayed-type hypersensitivity [DTH], and splenic lymphocyte proliferation), humoral immunity (serum hemolytic activity in immunized animals), and nonspecific immunity (peritoneal macrophages phagocytsis natural killer [NK] cell activity) were measured. Both *Bifidobacterium* strains independently increased the DTH response. Macrophage phagocytosis was also enhanced, while activities of the NK cells and levels of the serum hemolysin also were significantly higher than in the control group. There was a significant increase in splenic lymphocyte proliferation in bifidobacteria treatment animals compared to controls. In conclusion, ingestion of *Bifidobacterium adolescentis* BBMN23 and *Bifidobacterium longum* BBMN68 can enhance both innate and acquired immunity in healthy specific pathogen-free (SPF) mice and strains of Bifidobacteria from healthy centenarians in Bama longevity villages in China may possess potentially valuable immunomodulatory properties.

Martínez-Cañavate et al., (2009) evaluated the immunological effects of the consumption of a dairy product containing two probiotic strains in children suffering from allergy [157]. A double-blinded, randomized, control comparative study was performed with 44 allergic children. Children were randomly distributed in two groups, a control Yogurt and a Probiotic group. Both groups daily consumed 200 mL of a dairy fermented product for 3 months. The Yogurt group consumed a conventional yogurt, whereas the Probiotic group consumed a similar dairy product where *Lactobacillus bulgaricus* was substituted by a mixture of *Lactobacillus gasseri* CECT5714 and *Lactobacillus coryniformis* CECT5711 (at least 10^6 CFU/g each strain). Intestinal and immunological parameters were measured in fecal and blood samples. The consumption of the probiotic product induced a significant decrease in the level of IgE in plasma (p = 0.03) and an increase in $CD4^+/CD25^+$ T regulatory cells (p = 0.01). The decrease in IgE was accompanied by a significant increase in mucosal IgA (p = 0.01). However, changes in other effector cells potentially involved in allergic reactions such as eosinophiles, basophiles or other IgE^+ cells were not detected. The consumption of the probiotic product also induced significant changes in innate response as a significant increase in natural killer cells was detected (p = 0.03). The daily consumption of a probiotic product containing *Lactobacillus gasseri* CECT5714 and *Lactobacillus coryniformis* CECT5711 for 3 months induced, in allergic children, beneficial effects on immune parameters involved in the allergic response such as a reduction of IgE in plasma and an increase in regulatory T cells. The probiotic product also enhanced innate and specific immune parameters that may improve the general health status of children.

Bickert et al., (2009) investigated whether the application of the probiotic *Escherichia coli* Nissle 1917 (EcN) could protect mice from developing ovalbumin (OVA)-specific T helper-2 responses in the airways [158]. OVA-specific T_H2 responses were induced by 2 intraperitoneal (i.p.) injections with OVA/alum followed by 1 intranasal (i.n.) challenge with OVA. EcN was given orally during the entire sensitization and challenge period, together with OVA/alum during the i.p. sensitizations, or i.n. before or during the airway challenge with OVA. When the bacteria were given together with OVA/alum airway eosinophilia, airway hyper-reactivity, goblet cell metaplasia and IL-5 levels in the bronchoalveolar lavage and mediastinal lymph node cell cultures were reduced. This effect was associated with increased numbers of IFN-gamma producing T helper-1 cells and IFN-γ levels in the airways and strongly increased OVA-specific IgG(2a) titers in the serum. The suppressive effect on airway eosinophilia was dependent on IFN-gamma but not TLR-4. Applying EcN i.n. or orally did not reduce the development of allergen-specific T_H2 responses. EcN can inhibit the development of allergic responses when the bacteria are present at the site of T_H2 cell priming and that this immunomodulatory effect is due to a shift from T_H2 to T_H1 response. The data support the hypothesis that probiotics may help reduce allergic responses and that EcN may also be used as adjuvant therapy to induce allergen-specific T_H1 responses.

Wang et al., (2009) determined the effect of feeding *Lactobacillus fermentum* I5007 on the immune system of weaned pigs with or without *Escherichia coli* challenge [159]. Twenty-four weaned barrows (6.07 ± 0.63 kg BW) were randomly assigned to one of four treatments (n = 6) in a factorial design experiment. The first two treatments consisted of healthy piglets with half of the pigs receiving no treatment while the other half was orally administered with *Lactobacillus fermentum* I5007 (10^8 CFU/mL) at a daily dose of 20 mL. Pigs in the second two treatments were challenged on the first day with 20 mL of *Escherichia coli* K88ac (10^8 CFU/mL). Half of these pigs were not treated while the remaining pigs were treated with 20 mL of *Lactobacillus fermentum* I5007 (10^8 CFU/mL). Peripheral blood lymphocytes subsets were determined using flow cytometry. The intestinal mucosal immunity of the pigs was monitored by real time polymerase chain reaction. The cytokine content of the pig's serum was also analyzed. Oral administration of *Lactobacillus fermentum* I5007 increased blood $CD4^+$ lymphocyte subset percentage as well as TNF-α and INF-γ expression in the ileum. Pigs challenged with *Escherichia coli* had elevated jejunal TNF-α while interferon-c expression was increased throughout the small intestine. There was no difference in the concentration of the cytokines IL-2, IL-6, TNF-α and INF-γ in the serum. $CD8^+$ and $CD4^+/CD8^+$ in peripheral blood were not affected by treatment. *Lactobacillus fermentum* I5007 enhanced T cell differentiation and induced ileum cytokine expression suggesting that this probiotic strain could modulate immune function in piglets.

Ouwehand et al., (2009) aimed at modifying the intestinal microbiota with a potential synbiotic on selected immune and microbiota markers [160]. Healthy elderly subjects were randomized to consume during 2 weeks either a placebo (sucrose) or a combination of lactitol and *Lactobacillus acidophilus* NCFM twice daily in a double-blind parallel trial. After the intervention, stool frequency was higher in the synbiotic group than in the placebo group and a significant increase in faecal *Lactobacillus acidophilus* NCFM levels was observed in the synbiotic group, after baseline correction. The study subjects had faecal *Bifidobacterium* levels that were similar to those reported in healthy young adults. These levels were, nevertheless, significantly increased by the intervention. Levels of SCFA were not changed significantly. Of the measured immune markers, PGE_2 levels were different between treatments and IgA levels changed over time. These changes were modest which may relate to the fact that the volunteers were healthy. Spermidine levels changed over time which may suggest an improved mucosal integrity and intestinal motility. Consumption of lactitol combined with *Lactobacillus acidophilus* NCFM twice daily could improve some markers of the intestinal microbiota composition and mucosal functions.

Baron (2009) conducted a controlled study to evaluate the effects of the patented GanedenBC30 probiotic (*Bacillus coagulans* GBI-30, 6086, marketed as Sustenex [Ganeden Biotech, Inc., Mayfield Heights, OH]) on the immune system when exposed to adenovirus and influenza in otherwise healthy adults [161]. Ten healthy men and women (average age, 44 years) were instructed to consume 1 capsule of GanedenBC30 with water once a day for 30 days. At baseline and after completion of the 30-day treatment, blood levels of cytokines were measured *in vitro* after T-cell exposure to adenovirus and influenza A. Each participant served as his/her own control with baseline blood draw. The use of GanedenBC30 significantly increased T-cell production of TNF-α in response to adenovirus exposure ($p = 0.027$) and influenza A (H3N2 Texas strain) exposure ($p = 0.004$), but it did not have a significant effect on the response to other strains of influenza. No serious adverse events were reported throughout the study. The patented GanedenBC30 probiotic could be a safe and effective therapeutic option for enhancing T-cell response to certain viral respiratory tract infections.

Tseng et al., (2009) evaluated the effects of *Bacillus subtilis* E20 isolated from fermented soybean on immune parameters and the disease resistance of the white shrimp (*Litopenaeus vannamei*) after 98 days of *Bacillus subtilis* E20 feeding [162]. Shrimp fed *Bacillus subtilis* E20-containing diets at concentrations of 10^6 (E206), 10^7 (E207), and 10^8 (E208) CFU/kg, respectively, had significantly increased survival rates of 13.3%, 16.7%, and 20%, compared to the control (fed no probiotic) after being challenged with *Vibrio alginolyticus*. There were no significant differences in the total hemocyte count, respiratory burst, or superoxide dismutase glutathione peroxidase among all treatments. Shrimp fed a higher concentration of the probiotic (E208) exhibited significant increases in phenoloxidase activity, phagocytic activity, and clearance efficiency compared to control shrimp. In addition, *Bacillus subtilis* E20 showed a weaker inhibitory effect against the growth of Aeromona hydrophila with around a 0.3-cm inhibitory zone, but showed no inhibitory effects against other selected pathogens, such as white shrimp pathogens: *Vibrio alginolyticus* and *Vibrio vulnificus*. The increased resistance of shrimp after *Bacillus subtilis* E20 consumption occurred through immune modifications, such as increases in phenoloxidase activity, phagocytic activity, and clearance efficiency against *Vibrio alginolyticus*.

Cammarota et al., (2009) analyzed the probiotic potential of *Lactobacillus plantarum* DSMZ 12028 *in vitro* using the pathogen *Escherichia coli* K4 and a certified probiotic, *Lactobacillus paracasei* F19, as controls [163]. Adhesion to intestinal epithelial cells was evaluated using two cell lines, CaCO-2 and HT-29, through the plate dilution method (figure 62). Moreover, the bacteria/epithelial dynamic interaction was continuously monitored using time-lapse microscopy. Expression of the innate immunity receptors, the TLRs, was evaluated by semi-quantitative PCR on an epithelial/bacteria coculture. Real-time PCR was used to monitor expression of TLRs and cytokines in a monocytic cell line (THP-1) following bacterial exposure. The adherence of the strain to intestinal epithelial cells was comparable to that of the probiotic. Time-lapse experiments showed that *Escherichia coli* K4 induced cell death while *Lactobacillus plantarum* did not affect proliferation at a 10:1 bacteria/cell ratio. *Lactobacillus plantarum* down-regulated TLR mRNAs with the exception of TLR2, while *Lactobacillus paracasei* F19 and *Escherichia coli* K4 caused a significant ($p < 0.05$) up-regulation of TLR2 and 4, respectively. To simulate the activation of immune cells in the lamina propria, the immunomodulation of *Lactobacillus plantarum* on a monocytic cell line THP-1 was analyzed. Proinflammatory cytokines, such as TNFα, were increased by the presence of bacteria.

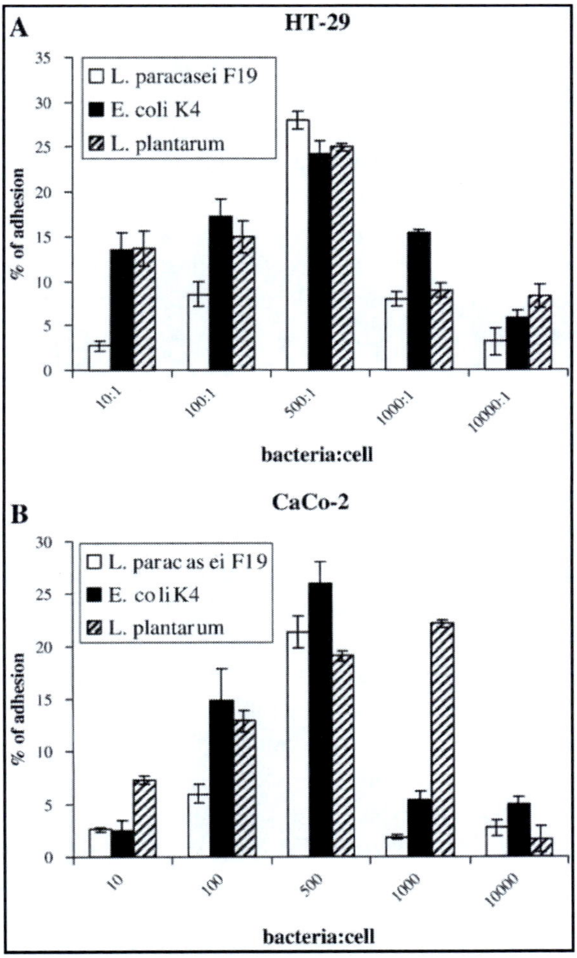

Figure 62. Adhesion assay of *Lactobacillus paracasei* F19, *Escherichia coli* K4 and *Lactobacillus plantarum* adhesion on HT-29 (panel A) and CaCo-2 cells (panel B). Data are expressed as the number of adherent bacteria relative to the number of seeded bacteria (% adhesion) [163].

The pathogen *Escherichia coli* K4 also induced a strong up-regulation of proinflammatory cytokines, such as IL8, IL1β and IL23. No differences were observed between experimental groups for IFNγ, IL-10 and IL12p40. *Lactobacillus plantarum* DSMZ 12028 demonstrated probiotic traits, inducing a proinflammatory response just above the "threshold level", which could prevent an inflammatory outcome, while inducing a higher state of alertness in the defense system of the host intestinal epithelial cells.

Lombardo et al., (2009) evaluated, in an open, short-term study, the efficacy and tolerability of a probiotic containing *Lactobacillus paracasei* subsp. *paracasei* F19 in irritable bowel syndrome (IBS) [164]. 100 patients with IBS (38 male, 62 fernale, age range I 8-68 years), 52 with diarrhoea and 48 with constipation, were recruited in accordance with the Rome II diagnostic criteria. All patients were administered Genefilus Fl9 (Siffra Pharmaceuticals SpA, Florence, Italy), a commercial preparation containing *Lactobacillus paracasei* subsp. *paracasei* F19, at a dose of one sachet (dissolved in water) twice daily for 14 days. Each sachet contained 12 x 10^9 genetically stable microorganisms of *Lactobacillus paracasei* subsp. *paracasei* F19, combined with 750 mg gluco-oligosaccharides plus vitamins Bl, B5 and B6. Symptoms were evaluated, based upon a daily diary, before and after treatment. In addition, in a series of 20 consecutive patients, *Lactobacillus paracasei* subsp. *paracasei* F19 content was evaluated in faeces before and after treatment. Following treatment, abdominal pain was no longer present in 94% of the patients with IBS with diarrhoea or in 87 % of those with IBS with constipation. Diarrhoea was no longer present in 88% of the patients, while constipation was no longer present in 83% of the patients. Abdominal distension had disappeared, or considerably improved, in 95% of the study population. Percentages remained unchanged at follow-up, 2 weeks after the end of treatment. No significant side effects or drop-outs were recorded. The results of microbiological evaluations performed on faeces revealed a low microbial *Lactobacillus paracasei* subsp. *paracasei* F19 load before treatment and a marked increase in load following treatment, with a minimum of 2, to a maximum of 10, CFU/g faeces. Genefilus F19 was confirmed to be efficacious and well tolerated in patients with IBS presenting with diarrhoea or with constipation and, therefore, useful in everyday clinical practice. Double-blind controlled studies, with placebo, including clinical evaluations at medium- and longterm follow-up, are still under way with a larger series of patients to confirm these promising findings.

Nishimura et al., (2009) performed a randomized, double-blind, placebo controlled study in 45 subjects with perennial allergic rhinitis (PAR) treated by oral administration of *Tetragenococcus halophilus* Th221 (high dose, 60 mg day, 15 subjects; low dose, 20.4 mg day, 15 subjects) or a placebo (15 subjects) for 8 weeks in order to evaluate its anti-allergic effects [165]. There were no significant differences among the groups that ingested Th221 and the placebo group regarding the disease severities, total nasal symptom scores and total nasal sign scores examined by physicians. However, the disease severity examined by physicians significantly improved in the high-dose group at the end of the trial compared with the beginning ($p < 0.05$). The total score for nasal symptoms of subjects who received a high dose of Th221 also showed a significant improvement at the end of the trial compared with the beginning ($p < 0.01$). According to the subjects' diaries, significant improvements in sneezing and rhinorrhea were observed during some periods in the high-dose group. The change in serum total immunoglobulin E (fig 63) improved significantly at the end of the trial compared with the beginning in this group ($p < 0.05$). The safety Th221 treatment was confirmed by laboratory tests and inspection of the general condition of each subject. Thus Th221 could be expected to safely improve the symptoms of PAR.

	Group	Week 0	Week 4	Week 8
Total IgE (IU/mL)	Placebo	381±130	372±117	343±101
	Low dose	377±73	391±86	358±71
	High dose	1122±745	1021±643	1021±643
Change in total IgE (%)	Placebo	100.0	105.4±6.5	101.1±3.8
	Low dose	100.0	100.7±3.5	95.0±3.0
	High dose	100.0	100.9±2.6	94.1±2.1
House dust specific IgE (Class)	Placebo	2.5±0.3	2.5±0.3	2.5±0.3
	Low dose	2.7±0.3	2.7±0.3	2.7±0.3
	High dose	3.1±0.3	3.1±0.3	3.1±0.3
D. pteronyssinus-specific IgE (Class)	Placebo	2.6±0.3	2.5±0.4	2.7±0.3
	Low dose	2.7±0.3	2.7±0.3	2.7±0.3
	High dose	3.0±0.4	3.0±0.4	3.1±0.4
Eosinophil count (%)	Placebo	5.5±0.9	4.7±0.8	5.2±1.0
	Low dose	4.7±0.8	4.2±0.5	4.9±0.8
	High dose	5.1±0.8	4.7±0.7	5.3±0.8

Figure 63. Effects of Th221 treatment on serum IgE and eosinophil counts of perennial allergic rhinitis subjects (Modified from Nishimura et al., 2009) [165].

Arribas et al., (2009) administered the probiotic strain *Escherichia coli* Nissle 1917 to rats or mice for 2 weeks in two experimental models of altered immune response, the trinitrobenzenesulphonic acid (TNBS) model of rat colitis, localized in the colon, and the lipopolysaccharide (LPS) model of systemic septic shock in mice [166]. Inflammatory status was evaluated both macroscopically and biochemically after 1 week in the TNBS model or after 24 h in the LPS shock model. In addition, splenocytes were obtained from mice and stimulated, *ex vivo*, with concanavalin A or LPS to activate T or B cells, respectively, and cytokine production (IL-2, IL-5 and IL-10) by T cells (figure 64) and IgG secretion by B cells measured. *Escherichia coli* Nissle

1917 was anti-inflammatory in both models of altered immune response. This included a reduction in the pro-inflammatory cytokine tumour necrosis factor-a, both in the intestine from colitic rats, and in plasma and lungs in mice treated with LPS. The systemic beneficial effect was associated with inhibited production of the T cell cytokines and by down-regulation of IgG release from splenocyte-derived B cells. Thus the anti-inflammatory effects of *Escherichia coli* Nissle 1917 given orally were not restricted to the gastrointestinal tract.

Group (n = 10)	TNF-α (pg·g⁻¹ tissue)	IL-10 (pg·g⁻¹ tissue)	LTB4 (pg·g⁻¹ tissue)
Non-colitic	365 ± 9	937 ± 99	2.6 ± 1.8
TNBS control	1322 ± 124	418 ± 56	9.9 ± 1.8
TNBS probiotic	870 ± 93*	516 ± 14	10.2 ± 0.8

Figure 64. Effects of treatment with Escherichia coli Nissle 1917 (1 x 109 CFU/d per rat) on cytokine [TNF-α and IL-10] and LTB4 production in trinitrobenzenesulphonic acid (TNBS) experimental colitis in rats [166].

In 2009, Jain et al. evaluated the feeding effect of probiotic dahi containing *Lactobacillus casei* on immune system in terms of cytokine gene expression in the spleen and Peyer's patches of mice [167]. Animals were divided into three groups and fed with; synthetic diet [control group (CD)], dahi containing mixed dahi culture [control dahi-fed group (CDF)]; and probiotic dahi fed group (PDF) for 28 days. The mRNA levels of IL-2, IL-4, IL-6 and IFN-γ were examined after 14 and 28 days. Total lactobacilli and lactococci loads were determined in the feces. The mRNA levels of IFN-γ in both spleen and Peyer's patches was found to be significantly increased in PDF animals after 14 and 28 days ($p < 0.05$) compared with CD and CDF groups.

The abundance of IL-2 mRNA also increased significantly in the Peyer's patches of PDF-fed animals. No significant changes were observed in mRNA levels of IL-4 and IL-6 in both spleen and Peyer's patches during whole experimental period. Further, total fecal lactobacilli and lactococci counts in the PDF group were significantly increased during first 10 days, then remained higher up to day 28 compared to other two groups. It was concluded that feeding probiotic dahi enhanced the expression of T_H1 type cytokines (IFN-γ and IL-2), especially in the mucosal immune organ (Peyer's patches) rather than in systemic organs (the spleen).

This indicated that feeding with probiotic dahi could strengthen the host immune system and protect against the progression of various immune-mediated diseases.

West et al. in 2009 evaluated the effects of feeding *Lactobacillus* F19 during weaning on the incidence of eczema and T_H1/T_H2 balance(168). In a double-blind, placebo-controlled randomized intervention trial, infants were fed cereals with (n = 89) or without *Lactobacillus* F19 (n = 90) from 4 to 13 months of age. The cumulative incidence of eczema at 13 months was assessed. The ratio of INF-γ to IL 4 mRNA expression levels in polyclonally stimulated peripheral blood T cells was used as a proxy for immune balance. Total and specific IgE serum levels were also assessed. The cumulative incidence of eczema at 13 months was 11% (4-17%, 95% CI) and 22% (13-31%, 95% CI) in the probiotic and placebo groups, respectively ($p < 0.05$). The number needed to treat was 9 (6.5-11.5, 95% CI). At 13 months of age, the INF-γ/IL4 mRNA ratio was higher in the probiotic compared with the placebo group ($p < 0.05$).

In contrast, there were no differences between groups in serum concentration of IgE. Feeding *Lactobacillus* F19 during weaning could be an effective tool in the prevention of early manifestation of allergy, e.g., eczema. The higher T_H1/T_H2 ratio in the probiotic compared with the placebo group suggested enhancing effects of *Lactobacillus* F19 on T cell-mediated immune response.

Vlieger et al., (2009) randomly administered, to 126 newborns, a prebiotic-containing starter formula supplemented with *Lactobacillus paracasei* ssp. *paracasei* and *Bifidobacterium animalis* ssp. *lactis* or the same formula without probiotics for the first 3 months of life [169]. A total of eighty infants continued the study until they were aged 6 months. Growth measurements were taken monthly at healthy baby clinics. Diaries were used to monitor behavior, infections, use of antibiotics, as well as stool characteristics. Normal growth occurred in all infants and no statistically significant differences were detected between the probiotics group and the control group for gain in weight, length and head circumference. Infants in the probiotics group produced softer and more frequent stools during the first three months of life. No differences were found in crying and sleeping hours, number of parent-diagnosed infections, antibiotic use, visits to the general practitioner and number of adverse events (figure 65). The use of a prebiotic-containing starter formula supplemented with *Lactobacillus paracasei* ssp. paracasei *Bifidobacterium animalis* ssp. lactis in early infancy was safe, well tolerated and had no adverse effects on growth and infant behavior.

	1th trimester Probiotics (n 53)		Control (n 47)		2nd trimester Probiotics (n 41)		Control (n 38)	
	Mean	SD	Mean	SD	Mean	SD	Mean	SD
Crying (h/d)	1.8	1.0	1.8	0.9	0.9	0.7	1.0	0.9
Sleeping (h/d)	15.3	1.5	15.2	1.1	15.0	1.2	14.6	1.3
Upper respiratory tract infection (times/month)	0.80	0.77	0.75	1.0	1.1	0.8	1.0	1.1
Gastrointestinal infestion (times/month)	0.27	0.74	0.13	0.34	0.18	0.33	0.36	0.75
Visist to to general practitioner (times/month)	0.25	0.42	0.27	0.33	0.28	0.50	0.29	0.55
Use of antibiotics (times/month)	0.044	0.15	0.082	0.19	0.069	0.18	0.079	0.29

Figure 65. Parameters of general health during the first and second trimester (Modified from Vlieger et al., 2009) [169].

Chapter IV

Interesting Perspectives and Insights on Probiotics

Simrén et al., (2010) studied the effects of a probiotic product on IBS symptoms. 74 IBS patients were randomized to receive 8 weeks of daily treatment with 400 mL milk fermented with the yoghurt bacteria and containing *Lactobacillus paracasei* subsp. *paracasei* F19, *Lactobacillus acidophilus* La5 and *Bifidobacterium lactis* Bb12 (Cultura; active) or acidified milk without these bacteria (control) [170]. The primary endpoint was the proportion of subjects reporting adequate relief of their IBS symptoms at least 50% of the weeks. IBS symptom severity, psychological symptoms and quality of life were assessed. The proportion of responders was 38% (14/37 patients) in the active group and 27% (10/37 patients) in the control group (p = 0.3). IBS symptom severity improved significantly in both groups during the treatment period. This change was greater in the active group during the first 2 weeks, but thereafter, no significant group differences were seen. A clearly positive effect of fermented milk containing three probiotic bacteria on GI symptoms in IBS patients was not detectable when compared with the control treatment. However, a trend towards a more favourable effect during the first weeks was seen in the active group.

Sierra et al., (2010) evaluated the tolerance and beneficial effects in healthy adults of the strain, *Lactobacillus salivarius* CECT5713 isolated from breast milk [171]. A phase II, randomized, double-blinded, placebo-controlled human clinical trial was carried out in 40 healthy adults. The Probiotic group received a daily dose of 2 x 10^8 CFU of *Lactobacillus salivarius* CECT5713 in capsules during 4 weeks while volunteers of the control received only a placebo. Gastrointestinal and immunological parameters were analyzed. *Lactobacillus salivarius* CECT5713 was well tolerated and no adverse effects were detected. Consumption of the probiotic strain increased fecal lactobacilli counts (7.9+/-0.1 vs. 7.05+/-0.2 CFU/g feces, $p = 0.001$). Also, an improvement in the frequency of defecation ($p = 0.04$) was observed. Probiotic treatment induced significantly the percentage of NK cells and monocytes, as well as the plasmatic levels of immunoglobulins M, A and G, and the regulatory cytokine IL-10 (72.3+/-11.7 in probiotic group vs. 27.3+/-6.4 pg/mL in control group, $p < 0.01$). Daily administration of *Lactobacillus salivarius* CECT5713 to healthy adults was safe and improved gut microbiota and different parameters related to immune response.

Zhang et al., (2010) assess the effectiveness of *Lactobacillus rhamnosus* GG (LGG) in reducing the incidence and severity of post-weaning diarrhoea in piglets, and to investigate its mechanisms of action [172]. Eighteen weaned barrows were allocated to non-challenged control (NCN), challenged control (CCN) and LGG treatment (LGG) groups. Diarrhoea incidence was significantly lower in group LGG than group CCN after *Escherichia coli* challenge. Faecal coliform bacteria counts were significantly increased, while Lactobacilli and Bifidobacteria counts were decreased, in group CCN compared with the other groups after challenge. In the jejunum and ileum, sIgA concentrations were significantly higher in group LGG than in group CCN. In group LGG, administration of short-term LGG before *Escherichia coli* infection attenuated the elevation of serum IL-6 induced by *Escherichia coli*. Significantly higher concentrations of TNF-α were observed in group LGG than NCN and CCN at 6 h. IL-1β concentrations in group NCN were significantly higher than LGG at 6 hours and higher than CCN at 24 hours. In conclusion, LGG was effective in ameliorating diarrhoea in post-weaning piglets induced by *Escherichia coli* K88, possibly via modulation of intestinal microflora, enhancement of intestinal antibody defence, and regulation of production of systemic inflammatory cytokines.

Maldonado et al., (2010) evaluated the safety of a follow-on formula with *Lactobacillus salivarius* CECT5713 in 6-mo-old children [173]. The antibiotic susceptibility of *Lactobacillus salivarius* CECT5713 was analyzed by a dilution method. A double-blinded, randomized, placebo controlled study was performed. Children (n = 80) were distributed in two groups and consumed the formula supplemented or not with probiotics (2×10^6 CFU/g) during 6 months. Fecal samples were collected at enrollment, at 3 months, and at the end of trial. Clinical and anthropometric evaluations were performed. Depending on the variable, one-way or two-way repeated measures analysis of variance were used for the statistical analysis. The antibiotic susceptibility profile of the strain resulted as safe. No adverse effects associated with the consumption of the probiotic formula were reported. In addition, clinical parameters did not differ between groups. Consumption of the probiotic supplemented formula led to an increase in the fecal lactobacilli content (7.6 ± 0.2 versus 7.9 ± 0.1 log CFU/g, $p < 0.05$). *Lactobacillus salivarius* CECT5713 was detected in the feces of volunteers from the probiotic group. Probiotic consumption induced a significant increase in the fecal concentration of butyric acid at 6 months. A follow-on formula with *Lactobacillus salivarius* CECT5713 was safe and well tolerated in 6-months-old infant.

Zeuthen et al., (2010) isolated IEC (intestinal epithelial cells) from mouse foetuses at gestational day 19 and from germfree neonates, to mirror the influence of the very first bacterial stimuli such IEC [174]. These were stimulated with gut-derived bacteria, Gram-negative *Escherichia coli* Nissle and Gram-positive *Lactobacillus acidophilus* NCFM, and expression of genes important for immune regulation was measured together with cytokine production. *Escherichia coli* Nissle and Lactobacillus acidophilus NCFM strongly induced chemokines and cytokines, but with different kinetics, and only *Escherichia coli* Nissle induced down-regulation of Toll-like receptor 4 and up-regulation of Toll-like receptor 2. The sensitivity to stimulation was similar before and after birth in germ-free IEC, although Toll-like receptor 2 expression was higher before birth than immediately after. In conclusion, IEC isolated before gut colonization occurs at birth, are highly responsive to stimulation with gut commensals, with *Lactobacillus acidophilus* NCFM inducing a slower, but more sustained response than *Escherichia coli* Nissle. *Escherichia coli* could induce intestinal tolerance through very rapid up-regulation of chemokine and cytokine genes and down-regulation of Toll-like receptor 4, while regulating also responsiveness to Gram-positive bacteria.

Generoso et al., (2010) assessed whether oral treatment with viable or heat-killed yeast *Saccharomyces cerevisiae* strain UFMG 905 could prevent bacterial translocation (BT), intestinal barrier integrity, and stimulate the immunity, in a murine intestinal obstruction (IO) model [175]. Four groups of mice were used: mice undergoing only laparotomy (CTL), undergoing intestinal obstruction (IO) and undergoing intestinal obstruction after previous treatment with viable or heat-killed yeast. BT, determined as uptake of 99m Tc-*Escherichia coli* in blood, mesenteric lymph nodes, liver, spleen and lungs, was significantly higher in IO group than in CTL group. Treatments with both yeasts reduced BT in blood and all organs investigated. The treatment with both yeasts also reduced intestinal permeability as determined by blood uptake of 99m Tc-DTPA. Immunological data demonstrated that both treatments were able to significantly increase IL-10 levels, but only viable yeast had the same effect on sIgA levels. Intestinal lesions were more severe in IO group when compared to CTL and yeasts groups. Thus both viable and heat-killed cells of yeast prevented BT, probably by immunomodulation and by maintaining gut barrier integrity. Only the stimulation of IgA production seemed to depend on the yeast viability.

Foligné et al., (2010) evaluated the *in vitro* immunomodulation capacity of various non-pathogenic yeast strains and to investigate the ability of some of these food grade yeasts to prevent experimental colitis in mice [176]. *In vitro* immunomodulation was assessed by measuring cytokines (IL-12p70, IL-10, TNF and INF-γ) released by human peripheral blood mononuclear cells (figure 66) after 24 hours stimulation with 6 live yeast strains (*Saccharomyces* spp.) and with bacterial reference strains. A murine model of acute 2-4-6-trinitrobenzene sulfonic acid (TNBS)-colitis was next used to evaluate the distinct prophylactic protective capacities of three yeast strains compared with the performance of prednisolone treatment.

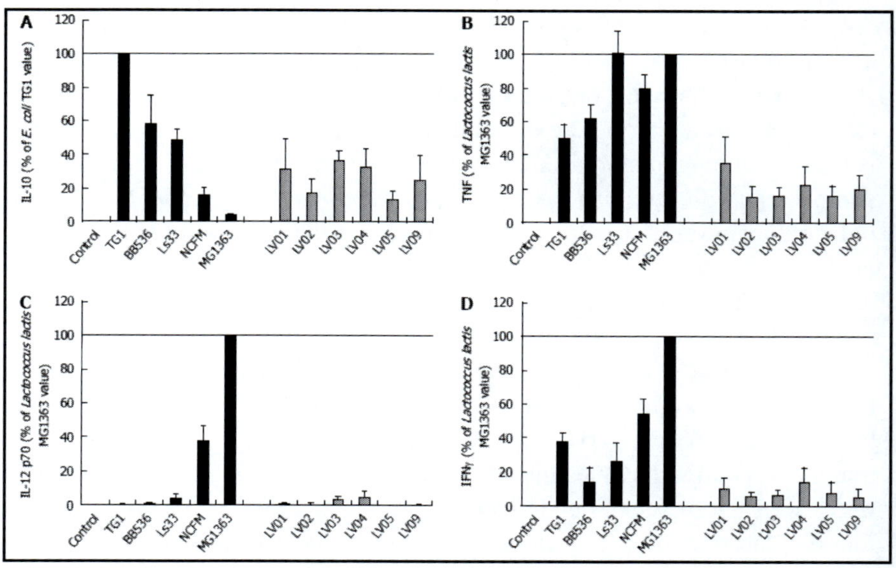

Figure 66. Cytokine production in human PBMCs, stimulated with bacterial reference strains and 6 yeast strains for 24 h. Cytokine levels were analyzed in supernatants by ELISA. A: IL-10; B: TNF; C: IL-12 p70; D: IFN-γ. Results are expressed in % of *Escherichia coli* TG1 value for panel A and in % of *Lactococcus lactis* MG1363 for B, C and D, including mean ± SE; n = 5 [176].

The six yeast strains all showed similar non-discriminating anti-inflammatory potential when tested on immunocompetent cells *in vitro*. However, although they exhibited similar colonization patterns *in vivo*, some yeast strains showed significant anti-inflammatory activities in the TNBS-induced colitis model, whereas others had weaker or no preventive effect at all, as evidenced by colitis markers (body-weight loss, macroscopic and histological scores, myeloperoxidase activities and blood inflammatory markers).

Hemmerling et al., (2010) in a phase 2a study assessed colonization efficiency, safety, tolerability, and acceptability of *Lactobacillus crispatus* CTV-05 (Lactin-V) administered by a vaginal applicator [177]. Twenty-four women with BV were randomized in a 3:1 ratio of active product to placebo. Participants used Lactin-V at 2 × 10 CFU/dose or placebo for 5 initial consecutive days, followed by a weekly application over 2 weeks. They returned for follow-up on Days 10 and 28. Sixty-one percent of the 18 women randomized to the Lactin-V group were colonized with *Lactobacillus crispatus* CTV-05 at Day 10 or Day 28. Among LACTIN-V users with complete adherence to the study regimen, 78% were colonized at Day 10 or Day 28. Of the 120 adverse events (AEs) that occurred, 108 (90%) and 12 (10%) were of mild and moderate severity, respectively. AEs were evenly distributed between the Lactin-V and placebo group. Of the total AEs, 93 (78%) were genitourinary in origin. The most common genitourinary AEs included vaginal discharge (46%), abdominal pain (46%), dysuria (21%), pollakiuria (21%), vaginal odor (21%), and genital pruritus (17%). No grade 3 or 4 AEs or serious AEs occurred and no deep epithelial disruption was seen during colposcopic evaluation. The product was well tolerated and accepted.

Pérez-Cano et al., (2010) evaluated the potential of two *Lactobacillus* strains isolated from human milk to modulate the activation and cytokine profile of peripheral blood mononuclear cell (PBMC) subsets *in vitro* [178]. Moreover, these effects were compared to the same probiotic species of non-milk origin. *Lactobacillus salivarius* CECT5713 and *Lactobacillus fermentum* CECT5716 at 10^5, 10^6 and 10^7 bacteria/mL were co-cultured with PBMC (10^6/mL) from eight healthy donors for 24 hours. Activation status (CD69 and CD25 expressions) of natural killer (NK) cells ($CD56^+$), total T cells ($CD3^+$), cytotoxic Tcells ($CD8^+$) and $CD4^+$ T cells was determined by flow cytometry. Regulatory T cells (Treg) were also quantified by intracellular Foxp3 evaluation. Regarding innate immunity, NK cells were activated by addition of both *Lactobacillus* strains, and in particular, the $CD8^+$ NK subset was preferentially induced to highly express CD69 (~90%, $p < 0.05$). With respect to acquired immunity, approximately 9% of $CD8^+$ T cells became activated after co-cultivation with *Lactobacillus fermentum* or *Lactobacillus salivarius*. Although $CD4^+$ T cells demonstrated a weaker response, there was a preferential activation of Treg cells ($CD4^+CD25^+Foxp3^+$) after exposure to both milk probiotic bacteria ($p < 0.05$). Both strains significantly induced the production of a number of cytokines (figure 67) and chemokines, including TNF-α, IL-1β, IL-8, MIP-1α, MIP-1β, and GM-CSF, but some strain-specific effects were apparent. Thus *Lactobacillus salivarius* CECT5713 and *Lactobacillus fermentum* CECT5716 enhanced both natural and acquired immune responses, as evidenced by the activation of NK and T cell subsets and the expansion of Treg cells, as well as the induction of a broad array of cytokines.

O'Mahony et al., (2010) investigated the potential molecular and cellular mechanisms underpinning the *Bifidobacterium animalis* AHC7 protective effect [179]. Following 4 hours of infection with *Salmonella typhimurium*, NF-kB activation was significantly elevated *in vivo* in placebo and *Enterococcus faecium*-fed animals while *Bifidobacterium animalis* AHC7 consumption significantly attenuated the NF-kB response. *In vitro* anti-CD3/CD28 stimulated Peyer' s patch cells secreted significantly less TNF-α and IFN-γ following *Bifidobacterium animalis* AHC7 consumption (figure 68). Stimulated cells released more IL-12p70 but this difference did not reach statistical significance. No alteration in mucosal IL-6, IL-10 or MCP-1 levels were observed. No statistically significant change in the cytokine profile of mesenteric lymph node cells was noted. *In vitro*, *Bifidobacterium animalis* AHC7 was bound by dendritic cells and induced secretion of both IL-10 and IL-12p70. In addition, co-culture of $CD4^+$ T cells with *Bifidobacterium animalis* AHC7-stimulated dendritic cells resulted in a significant increase in $CD25^+Foxp3^+$ T cell numbers. *Bifidobacterium animalis* AHC7 exerts an anti-inflammatory effect via the attenuation of proinflammatory transcription factor activation in response to an infectious insult associated with modulation of pro-inflammatory cytokine production within the mucosa. The cellular mechanism underpinning *Bifidobacterium animalis* AHC7 mediated attenuation of NF-*k*B activation may include recognition of the bacterium by dendritic cells and induction of $CD25^+Foxp3^+$ T cells.

Chiu et al., (2010) calculated the percent weight gain (PWG) and feeding efficiency (FE) of *Epinephelus coioides* [180]. The survival of *Saccharomyces cerevisiae* P13 in the posterior intestines using a specific primer pair of YMR245w-F/YMR245w-R, non-specific immune parameters of grouper, and its susceptibility to *Streptococcus* sp. and an iridovirus were determined when the fish were fed diets containing *Saccharomyces cerevisiae* at 0 (control), 10^3, 10^5, or 10^7 CFU/kg for 4 weeks.

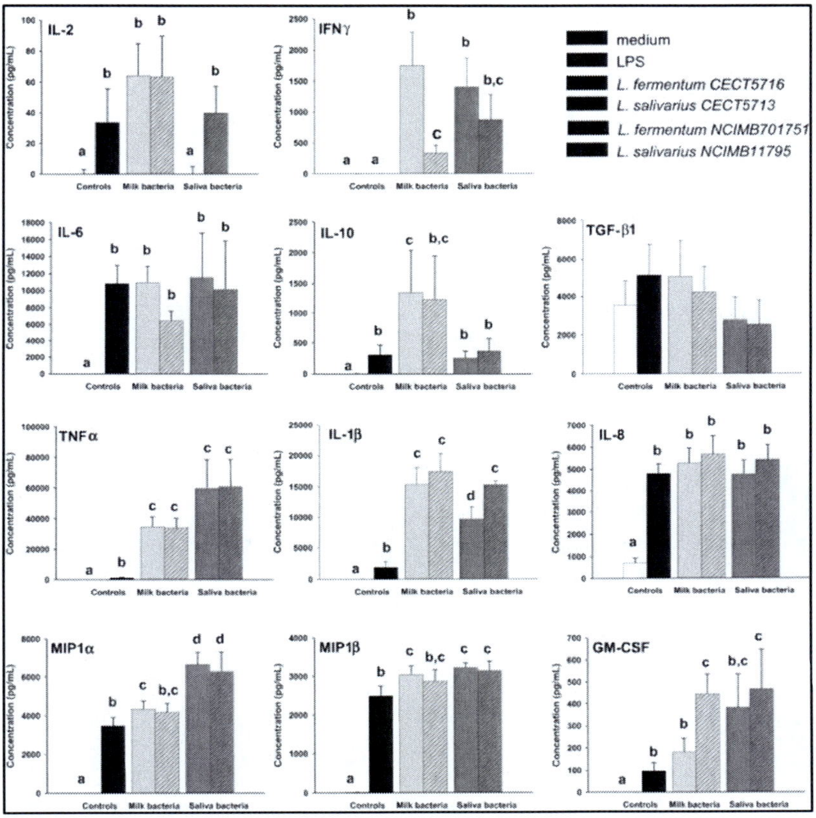

Figure 67. Production of cytokines by human PBMC in response to probiotic bacteria. Cytokines were analyzed by a flow cytometric multiple bead-based immunoassay or by ELISA in 24 h supernatants of PBMC cultures with bacteria: cell ratios of 1:1. Significant differences among control, lipopolysaccharide (LPS) and bacterial species were tested by one-way ANOVA. Means within agraphic without a common letter differ, $p < 0.05$ [178].

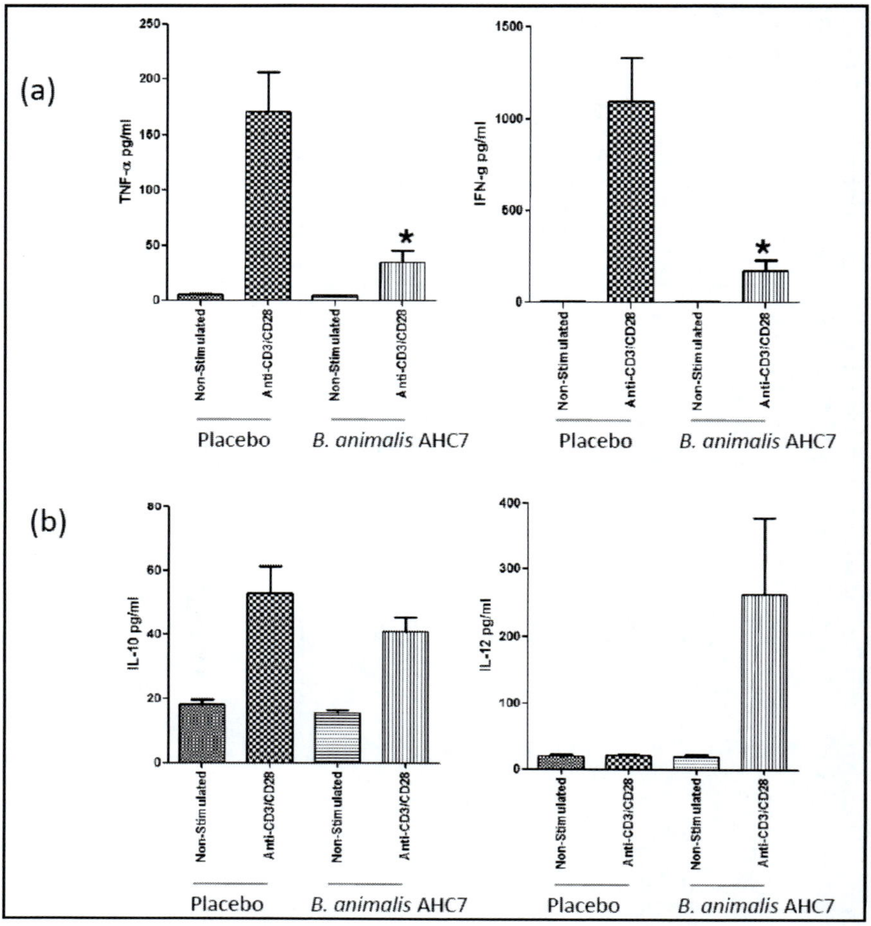

Figure 68. Peyer's patch cytokine secretion. Isolated Peyer's patch cells were cultured *in vitro* with or without anti-CD3/CD28 stimulation. Secretion of TNF-α and IFN-γ was significantly reduced in animals pre-fed with *Bifidobacterium animalis* AHC7 (a). Secretion of IL-10 was unaltered between the groups while a non-statistically significant increase in IL-12 secretion was observed (b). *$p < 0.05$ versus placebo (n = 10 animals per group) [179].

Results showed that grouper fed a diet containing *Saccharomyces cerevisiae* at the levels of 10^3, 10^5, and 10^7 CFU/kg had significantly increased PGW and FE especially in the 10^7 CFU/kg group which were 211.6% and 1.2, respectively.

Saccharomyces cerevisiae was able to survive in the fish posterior intestines during the *Saccharomyces cerevisiae* feeding period. Fish fed a diet containing *Saccharomyces cerevisiae* at 10^7 CFU/kg had significantly higher survival rates than those fed the 10^3 CFU/kg *Saccharomyces cerevisiae* diet and the control diet after challenge with *Streptococcus* sp. and an iridovirus, with increased survival rates of 26.6% and 36.6%, respectively, compared to the challenge control group. The phagocytic activity, respiratory burst and superoxide dismutase (SOD) level of head kidney leucocytes as well as serum lysozyme activity and serum alternative complement activity (ACH50) of fish fed diets containing *Saccharomyces cerevisiae* at 10^5 and 10^7 CFU/kg were significantly higher than those of fish fed the 10^3 CFU/kg *Saccharomyces cerevisiae*-contained diet and the control diets after 4 weeks of feeding, and had increased by 20% and 20%, 27.6% and 19.7%, 30.5% and 36.2%, 205.8% and 169.6%, and 90.8% and 80.3%, respectively, compared to the control group.

Thus dietary *Saccharomyces cerevisiae* administration of 10^5 and 10^7 CFU/kg to *Epinephelus coioides* was recommended to promote growth and enhance immunity and resistance against *Streptococcus* spp. and an iridovirus especially in the 10^7 CFU/kg group.

Allen et al., (2010) evaluated, in a randomized, double-blinded, placebo-controlled trial, the safety of a bacterial dietary supplement for the prevention of atopy in infants [181]. Two strains of lactobacilli (*Lactobacillus salivarius* CUL61 and *Lactobacillus paracasei* CUL08) and Bifidobacteria (*Bifidobacterium animalis* subsp. *lactis* CUL34 and *Bifidobacterium bifidum* CUL20) with a total of 1 x 10^{10} CFU were administered daily to women during the last month of pregnancy and to infants aged 0–6 months. Adverse events (AE) were classified according to WHO International Statistical Classification of Diseases criteria. Common symptoms were recorded by regular questionnaires.

Baseline characteristics of 220 mother-infant dyads in the treatment and 234 in the placebo group were similar. Compliance with the trial interventions, loss to follow-up, symptoms, drug usage, infant growth, method of feeding, visits to the doctor, and mothers' assessment of infant health were similar in the two groups. Fifteen (6.8%) mothers and 73 (33.2%) infants in the treatment group and 21 (9.0%) mothers and 75 (32.1%) infants in the placebo group reported AE ($p = 0.49$ and $p = 0.84$, respectively). Severe AE occurred in 18 mothers and 63 infants with a similar frequency in each group. None of the AE was attributed to the intervention. These organisms during pregnancy and early infancy were safe.

Gronbach et al., (2010) examined the role of the intestinal microbiota and the adaptive immune system in preventing translocation of probiotics such as *Escherichia coli* Nissle 1917 [182]. C57BL/6J mice raised under germfree conditions (GF-raised C57BL/6J mice) and Rag1$^{-/-}$ mice raised under germfree conditions (GF-raised Rag1$^{-/-}$ mice) and under specific-pathogen-free conditions (SPF-raised Rag1$^{-/-}$ mice) with probiotic *Escherichia coli* strain Nissle 1917, strain Nissle 1917 mutants, the commensal strain *Escherichia coli* mpk, or *Bacteroides vulgatus* mpk were challenged. Additionally, Rag1$^{-/-}$ mice with CD4^{+} T cells were reconstituted. *Escherichia coli* translocation and dissemination and the mortality of mice were assessed. In GF-raised Rag1$^{-/-}$ mice, but not in SPF-raised Rag1$^{-/-}$ mice or GF-raised C57BL/6J mice, oral challenge with *Escherichia coli* strain Nissle 1917, but not oral challenge with *Escherichia coli* mpk, resulted in translocation and dissemination. The mortality rate was significantly higher for *Escherichia coli* strain Nissle 1917-challenged GF-raised Rag1$^{-/-}$ mice (100%; $p < 0.001$) than for *Escherichia coli* strain Nissle 1917-challenged SPF-raised Rag1$^{-/-}$mice (0%) and GF-raised C57BL/6J mice (0%). Translocation of and mortality due to strain *Escherichia coli* Nissle 1917 in GF-raised Rag1$^{-/-}$ mice were prevented when mice were reconstituted with T cells prior to strain *Escherichia coli* Nissle 1917 challenge, but not when mice were reconstituted with T cells after *Escherichia coli* strain Nissle 1917 challenge. Co-colonization experiments revealed that *Escherichia coli* mpk could not prevent translocation of *Escherichia coli* Nissle 1917. Moreover, it was demonstrated that neither lipopolysaccharide structure nor flagella played a role in *Escherichia coli* strain Nissle 1917 translocation and dissemination. If both the microbiota and adaptive immunity are defective, translocation across the intestinal epithelium and dissemination of the probiotic *Escherichia coli* strain Nissle 1917 could occur and have potentially severe adverse effects. Harikrishnan et al., (2010) evaluated the effect of *Lactobacillus sakei* BK19 (10^8 cells/g) supplemented diet fed to kelp grouper, *Epinephelus bruneus* against streptococcosis caused by *Streptococcus iniae* and *Streptococcus parauberis* with reference to the innate immune response and disease resistance at 1, 2, and 4 weeks [183]. Maximum reduction in mortalities was observed in kelper feeding the probiotic diet for two weeks after challenged with the pathogens when compared to the infected group fed with basal diet; similarly the cellular and humoral immune responses such as head kidney macrophage phagocytic and peroxidase activities, serum lysozyme activity, and total protein levels increased significantly. In streptococcosis infected kelp grouper feeding *Lactobacillus sakei* BK19 enriched diet affords a higher level of disease protection due to stimulation of immune system. Kobayashi et al., (2010) evaluated the safety of two probiotic bacterial strains, *Lactobacillus casei* strain Shirota (LcS) and *Bifidobacterium breve* strain Yakult (BbY), these probiotics were orally administered to Lewis rats with experimental autoimmune encephalomyelitis (EAE), the experimental model of human multiple sclerosis [184]. Three experimental designs were examined by combining different antigen types and probiotic administration periods: (1) EAE was induced with a homogenate of guinea pig spinal cord as the sensitizing antigen, and LcS was orally administered from one week before this sensitization until the end of the experiment; (2) EAE was induced using guinea pig originated myelin basic protein (MBP) as the sensitizing antigen, and LcS was orally administered from one week before this sensitization to the end of the experiment; (3) EAE was induced using guinea pig MBP as the sensitizing antigen, and the probiotic strains (LcS and BbY) were administered starting in infancy (2 weeks old) and continued until the end of the experiment. In experiment 1, oral administration of LcS tended to suppress the development of neurological symptoms. Differences in neurological symptoms between the control group and the administration groups did not reach statistical significance in experiments 2 and 3. These results support the notion that neither LcS nor BbY exacerbates autoimmune disease.

Romanin et al., (2010) selected microorganisms from kefir able to down-regulate intestinal epithelial innate response and further characterize this activity [185]. CaCO-2 cells stably transfected with a human CCL20 promoter luciferase reporter were used to screen a collection of 24 yeast and 23 bacterial strains isolated from kefir. The Toll-like receptor 5 agonist, flagellin was used to activate the reporter cells, while pre-incubation with the selected strains was tested to identify strains with the capacity to inhibit cell activation. In this system, 21 yeast strains from the genera *Saccharomyces*, *Kluyveromyces* and *Issatchenkia* inhibited almost 100% of the flagellin-dependent activation, whereas only some lactobacilli strains showed a partial effect. *Kluyveromyces marxianus* CIDCA 8154 was selected for further characterization (figure 69). Inhibitory activity was confirmed at transcriptional level on CaCO-2/TC-7 and HT-29 cells upon flagellin stimulation. A similar effect was observed using other pro inflammatory stimulation such as IL-1β and TNF-α. Pre-incubation with yeasts induced a down-regulation of NF-κB signalling in epithelial cells *in vitro*, as well as expression of other pro-inflammatory chemokines such as CXCL8 and CXCL2. Furthermore, modulation of CCL20 mRNA expression upon flagellin stimulation was evidenced *in vivo*, in a mouse ligated intestinal loop model (figure 70).

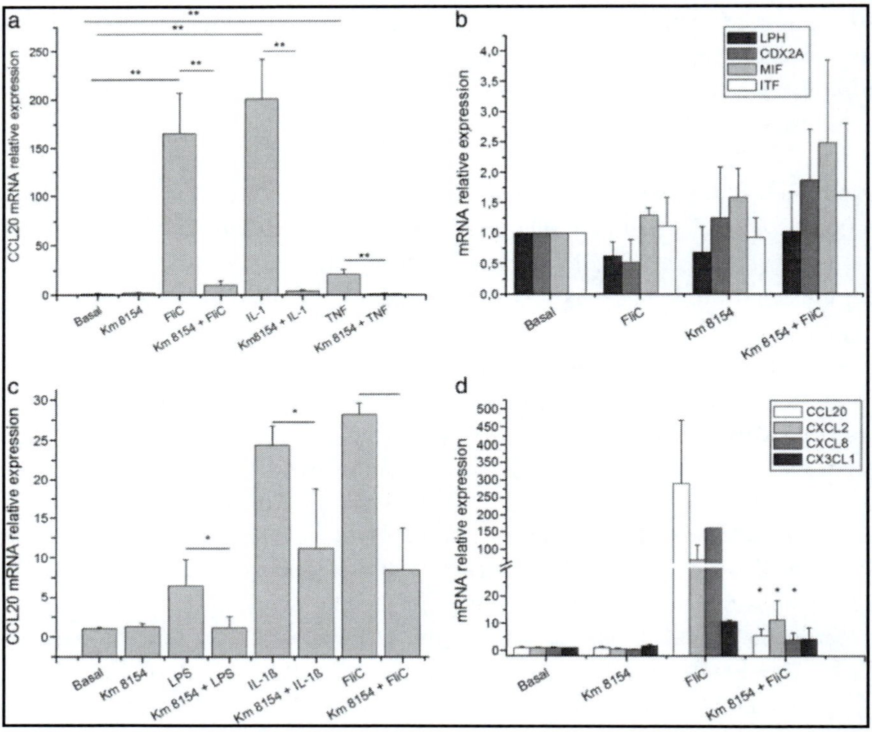

Figure 69. Effect of *Kluyveromyces marxianus* CIDCA 8154 (Km 8154) on: a. Expression of ccl20 mRNA on CaCO-2/Tc-7 cells (**$p < 0.01$). b. Gene expression levels related to different cell functions on CaCO-2/Tc-7 cells. c. Expression of ccl20 mRNA on HT-29 cells. d. Chemokine mRNA on CaCO-2/Tc-7 cells. Cells were incubated with Km 8154 and stimulated with flagellin (FliC), IL-1β or *Escherichia coli* purified LPS. (*$p < 0.05$). β-actin was used to normalize gene expression [185].

It was observed that kefir contained microorganisms able to abolish the intestinal epithelial inflammatory response that could explain some of the properties attributed to this fermented milk.

Immunoglobulin A (IgA) secretion in saliva decreased with age and could be the cause of increased vulnerability of the elderly to respiratory infections. Kotani et al., (2010) demonstrated the acceleration of salivary SIgA secretion by oral intake of *Lactobacillus pentosus* strain b240 (b240) in the elderly [186]. A total of 80 healthy elderly individuals were randomly allocated to either an intervention (i.e., b240) or a control (i.e., placebo) group. The elderly individuals in the b240 group were given a sterile water beverage (125 mL) containing heat-killed b240 (4×10^9 cells), while those in the placebo group were given only a sterile water beverage (125 mL); both groups received their respective beverages once daily for 12 weeks.

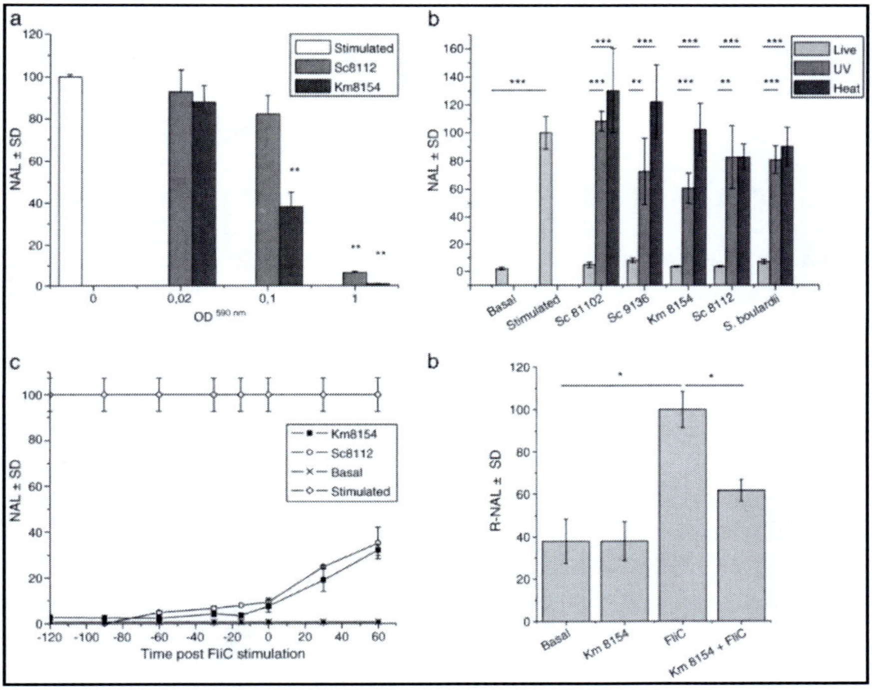

Figure 70. Effect of dose, viability and pre-incubation time of microorganisms over ccl20 expression. a. Effect of the dose of *Kluyveromyces marxianus* CIDCA 8154 (Km 8154) or *Saccharomyces cerevisiae* CIDCA 81102 (Sc 81102) on the CaCO2 ccl20:luc reporter system stimulated with flagellin. b. Effect of live (light gray), UV (gray) and heat-inactivated (dark gray) microorganism on the CaCO-2 ccl20:luc reporter system. c. Effect of pre-incubation time with kefir microorganisms on the CaCO-2 ccl20:luc. Data are representative of 3 independent experiments. d. Effect of incubation with Km 8154 on CaCO-2/Tc-7 cells transiently transfected with 3x-κB:luc and HSV-TK:renilla luciferase reporters. Renilla luciferase activity was used to normalize the transfection efficiency. Data are expressed as a percentage of the activity observed when flagellin alone was used as treatment (R-NAL), (*$p < 0.05$, **$p < 0.01$, ***$p < 0.001$) [185].

Saliva was collected before initiation of the study and every 2 weeks thereafter. Saliva flow rate and sIgA concentration were determined, and the sIgA secretion rate was calculated. The mean salivary sIgA secretion rate in the b240 group steadily increased until week 4 (exhibiting a 20% elevation relative to that at week 0), and then remained stable until week 12.

Changes in sIgA secretion rate over the intervention period were significantly greater in the b240 group than in the placebo group. The treatment groups exhibited no significant differences in adverse events. Oral intake of *Lactobacillus pentosus* strain b240 for 12 weeks significantly accelerated salivary sIgA secretion, thereby indicating its potential utility in the improvement of mucosal immunity and resistance against infection in the elderly.

Ligaarden et al., (2010) investigate the symptomatic effect of *Lactobacillus plantarum* MF1298 in subjects with IBS [187]. Primary outcome was treatment preference and secondary outcomes were number of weeks with satisfactory relief of symptoms and IBS sum score. The design was a randomised double blind placebo-controlled crossover trial. 16 subjects with IBS underwent two three-week periods of daily intake of one capsule of 10^{10} CFU *Lactobacillus plantarum* MF 1298 or placebo separated by a four-week washout period. Thirteen participants (81%; 95% CI 57% to 93%; p = 0.012) preferred placebo to *Lactobacillus plantarum* MF1298 treatment. The number of weeks with satisfactory relief of symptoms was statistically significantly higher in the placebo period compared with the *Lactobacillus plantarum* MF1298 period and the IBS sum score and the score for diarrhoea were significantly higher in the period of *Lactobacillus plantarum* MF1298 treatment than with placebo. This trial shows for the first time an unfavourable effect on symptoms in subjects with IBS after intake of a potential probiotic.

Won et al., (2011) tested some lactobacilli isolated from Kimchi, a Korean traditional food, for their capacity to modulate the T helper T_H1/T_H2 balance [188]. Ovalbumin (OVA)-sensitized mouse splenocytes were cultured with 26 strains of lactobacilli; the highest IL-12 induction and lowest IL-4 production were then observed in four strains, including *Lactobacillus plantarum* CJLP55, CJLP56, CJLP133, and CJLP136. These strains produced a larger amount of IL-12, which enhances differentiation and activation of T_H1 cells, in macrophage cell-lines more than positive control strains *Lactobacillus casei* KCTC 3109T and *Lactobacillus rhamnosus* GG, although they also induced production of IL-10, which is a suppressor of IL-12. Indeed, CJLP133-stimulated macrophages induced production of more T_H1 cytokine IFN-γ and less T_H2 cytokine IL-4 than KCTC 3109T and GG in co-cultivation with T cells. These findings suggest that lactobacilli from Kimchi may modulate the T_H1/T_H2 balance via macrophage activation in the hypersensitive reaction caused by T_H2 cells.

Mañé et al., (2011) aimed to investigate the effects of a mixture of two new probiotic strains of *Lactobacillus plantarum* CECT7315 and 7316 on systemic immunity in elderly [189]. 50 institutionalized elderly subjects were randomized, in a double-blind fashion, to receive for 12 weeks 1) $5 \cdot x\ 10^{8}$ CFU/day of *Lactobacillus plantarum* CECT7315/7316 ("low probiotic dose") (n = 13), 2) $5\ x \cdot 10^{9}$ CFU/day of the probiotic mixture ("high probiotic dose") (n = 19), or 3) placebo (n = 15). Leukocyte subpopulations, and cytokine levels (IL-1, IL-10, TGF-β1) were measured in venous blood at baseline, end of treatment (week 12), and end of follow-up (week 24). Infection and survival rates were recorded. After treatment, high probiotic dose resulted in significant increases in the percentages of activated potentially T-suppressor ($CD8^{+}CD25^{+}$) and NK ($CD56^{+}$ $CD16^{+}$) cells, while low probiotic dose increased activated T-helper lymphocytes ($CD4+CD25^{+}$), B lymphocytes ($CD19^{+}$), and antigen presenting cells (HLA-DR^{+}). Also, plasma TGF-β1 concentration significantly decreased after treatment with both probiotic doses. Most of these changes remained 12 weeks after probiotic discontinuation. Incidence of infections during treatment showed a significant trend to be lower in the high probiotic dose group. In addition, there was a significant trend for mortality to be greater in the placebo group vs. both probiotic groups. Depending on the dose, *Lactobacillus plantarum* CECT7315/7316 had different immune-enhancing effects in elderly subjects. These effects could result in a better clinical outcome.

Zhang et al., (2011) investigated the effects of porcine lactobacilli on the intestinal mucosal immunity of piglets [190]. Neonatal piglets were used as a model and were orally administrated with *Lactobacillus salivarius* B1 isolated from the duodenal mucosa of a healthy piglet. The feces of the piglets were collected on days 7, 14, and 21 for intestinal microflora analysis. On day 28, the piglets were sacrificed, and their intestinal mucosa samples were immediately collected to investigate the changes in intestinal morphological and immunocompetent cells. Finally, the expression of cytokines and TLRs was detected in the different intestinal segments. *Lactobacillus salivarius* B1 partially ameliorated the microflora of the feces and increased the number of intestinal immunocompetent cells, as the intraepithelial lymphocyte ($p < 0.05$), and the IgA-producing cells ($p < 0.01$) in the Lactobacilli-treated group were all increased compared with those in the control group. Enhanced expression of the cytokine IL-6 gene was also observed in the ileum ($p < 0.05$). Moreover, *Lactobacillus salivarius* B1 upregulated the expression of TLR2 in the intestinal tract at the gene and protein levels ($p < 0.05$). Thus *Lactobacillus salivarius* B1 was beneficial for the maturation of the intestinal mucosal immune system and elicited local immunomodulatory activities. In addition, the modulatory effects of *Lactobacillus salivarius* B1 on mucosal immunity mainly depended on its extracellular components.

Dos Santos et al., (2011) investigated the protective effects of *Lactobacillus delbrueckii* UFV-H2b20 on the resistance to *Listeria monocytogenes* infection in gnotobiotic mice [191]. Germfree mice or monoassociated mice were infected with Listeria monocytogenes, and the microbiological and immunological responses were evaluated after 1, 3, and 5 days of infection. Monoassociation with *Lactobacillus delbrueckii* was capable of protecting mice against death caused *by Listeria monocytogenes* and induced a faster clearance of the bacteria in the liver, spleen, and peritoneal cavity at days 1, 3, and 5 post-infection. Also, monoassociated mice displayed less liver injury than germfree mice. The production of TNF-α in the serum, peritoneal cavity, and gut was augmented in monoassociated mice. Likewise, the levels of IFN-γ found on supernatants of spleen cells cultures were higher after the monoassociation. In addition, increased production of nitric oxide in peritoneal cell cultures supernatants and in serum was observed in mice that received *Lactobacillus delbrueckii.* The monoassociation with *Lactobacillus delbrueckii* induced higher production of IL-10 in the mucosal immune system. Monoassociation with *Lactobacillus delbrueckii* UFV-H2b20 protected mice from death caused by *Listeria monocytogenes* infection by favoring effector responses while preventing their immunopathological consequences.

Manzoni et al., (2011) reported a 6-year, two-NICUs experience of routinary *Lactobacillus rhamnosus* GG (LGG) use in VLBW (very low birth weight) infants [192]. Clinical charts review, retrospective study of VLBW infants admitted to two Italian NICUs in the years 2003-2008. Standard protocol of LGG administration consisted of 3 x 10^9 CFU/day, in single oral dose, since 4th day-of-life, for 4-to-6-week courses. Nutritional policy relied on administration of fresh, expressed mother's milk, supplementation with preterm formula if needed. Data about LGG safety and tolerability, infections, feeding tolerance, microbiological clinical and surveillance cultures were retrieved and analyzed. Complete data were obtained for 743 of 811 VLBW infants. Mean birth-weight was 1056 g; mean gestational age 29.5 weeks. A total of 17.108 LGG doses were administered (mean 23.1/infant). No adverse effects or intolerances putatively attributable to LGG occurred. Overall, 5350 clinical and surveillance cultures from 13 different sites/devices were performed (mean: 7.2 cultures from 6.5 different sites/infant). None ever grew LGG, or other lactobacilli. No clinical sepsis episode was attributable to LGG.

Full enteral feeding was achieved at 19.2 mean days-of-life; 73% of infants were exclusively/partially breastfed. Fourteen NEC cases occurred (=1.9%), with 5 (=0.7%) being > 2b stage. Routinary supplementation of probiotic LGG in a large, 6-year VLBW infants Italian cohort proved microbiologically safe and clinically well tolerated.

Kingma et al., (2011) assessed the effect of *Lactobacillus johnsonii* (Ljo) N6.2 on mucosal inflammatory response using differentiated CaCO-2 monolayers [193]. The mRNA expression levels of CCL20, CXCL8, and CXCL10 chemokines were determined by qRT-PCR. Ljo strain at 10^{11} CFU/L induced a strong response in all chemokines examined. To assess the specific host-signaling pathways involved, RT-PCR amplification of Toll-like receptors (TLR) and nucleotide-binding oligomerization domainlike receptors was performed. TLR7 and TLR9 expression levels were induced 4.2- and 9-fold, respectively, whereas other TLR and nucleotide-binding oligomerization domain receptors were not modified. A similar effect was observed in CaCO-2 monolayers treated with Ljo cell-free extract or purified nucleic acids (NA). Increased levels of IFN type 1 and IFN regulators Stat1 and IRF7 followed the upregulation of TLR9. Activation of TLR9 was also evidenced by increased Frizzled 5 expression in Ljo-treated CaCO-2 cells and an increase in the number of Paneth cells in Ljo–fed, diabetes-prone rats. These results were in agreement with the polarizing-tolerizing mechanism described in which the apical stimulation of TLR9 in intestinal epithelial cells leads to a higher state of immunologic alertness.

Lee et al., (2011) examined 350 strains of lactic acid bacteria isolated from Korean infant feces and Kimchi [194]. Some strains were found to have immune modulatory and antimicrobial properties, while antagonistic activity against a panel of pathogenic bacteria was found to be strain dependent (figure 71). To evaluate the immune modulatory activity of the strains, lymphocyte IFN-γ secretion was determined in conjunction with cell proliferation. Some strains of *Lactobacillus gasseri*, *Lactobacillus fermentum* and *Lactobacillus plantarum* exhibited increased IFN-γ levels and lymphocyte proliferation. To evaluate the effects of these immune modulating lactobacilli on host life span, *Caenorhabditis elegans* was used as an *in vivo* model. Nematodes that were supplied heat-killed lactobacilli as a food source exhibited obvious differences in life span compared with those fed *Escherichia coli* OP50. The mean life span (determined as mean percent survival) of worms fed *Lactobacillus plantarum* CJLP133 and *Lactobacillus fermentum* LA12 was 13.89% and 13.69% greater, respectively, than that of control nematodes after 21 days (p = 0.036 and 0.043, respectively). In addition, some of safety profiles, including hemolytic type, gelatin hydration and degradation of urea, were found to be positive. These newly identified lactobacilli hold promise for use as probiotic agents, feed additives and/or in food applications.

Lactobacillus strains	Inhibition[a]					
	En. faecalis[b]	*E. coli*	*Staphy. aureus*	*Y. enterocolitica*	*S. typhimurium*	*L. monocytogenes*
Documented strains[c]						
L. rhamnosus GG	+	+	+	++	++	++
L. brevis KB290	−	−	−	−	−	−
L. plantarum L299v	+	+	+	++	+	−
Isolates from Korean infant feces						
L. gasseri CJLF1	+	+	+	+	+	+
L. gasseri CJLF6	−	−	−	−	−	−
L. gasseri CJLF20	+	+	−	++	++	−
L. gasseri CJLF21	+	+	−	++	++	−
L. gasseri CJLF22	+	+	−	++	++	−
L. gasseri CJLF23	+	+	+	+	+	−
L. gasseri CJLF34	+	+	+	+	+	+
L. gasseri CJLF35	+	+	+	++	+	−
L. gasseri CJLF36	−	−	−	−	−	−
L. gasseri CJLF37	−	−	−	−	−	−
L. gasseri CJLF48	−	−	−	−	−	−
L. gasseri CJMF3	+	+	+	+	+	−
L. gasseri CJMF15	−	−	−	−	−	−
L. gasseri CJMF18	−	−	−	−	−	−
L. gasseri CJMF19	−	−	−	−	−	−
L. gasseri CJMF29	−	−	−	−	−	−
L. gasseri CJMF31	−	−	−	−	−	−
L. gasseri CJMF36	−	−	+	+	−	−
L. gasseri CJMF41	−	−	+	+	−	−
L. fermentum LA12	+	+	+	++	++	+
L. fermentum CJMA3	−	+	+	++	+	+
L. fermentum CJMA7	−	−	−	+	−	++
L. fermentum CJMA9	−	−	−	+	+	+
L. fermentum CJMA10	−	−	−	+	−	+
L. fermentum CJMA11	−	−	−	++	+	+
L. plantarum CJMA1	++	++	++	++	++	++
L. plantarum CJMA2	−	+	−	+	+	++
Isolates from Korean Kimchi						
L. brevis D1	−	−	−	−	−	−
L. brevis D2	−	−	−	−	−	−
L. brevis D44	−	−	−	−	−	−
L. brevis CJD45	−	−	−	−	−	−
L. brevis CJD48	−	−	−	−	−	−
L. brevis CJD49	−	−	−	−	−	−
L. brevis CJD54	−	−	−	−	−	−
L. brevis CJD55	−	−	−	−	−	−
L. brevis CJD56	−	−	−	−	−	−
L. brevis CJNR27	−	−	−	−	+	−
L. brevis CJNR28	−	−	−	−	−	−
L. plantarum CJLP56	+	+	+	++	+	+
L. plantarum CJLP133	+	+	+	++	+	−
L. plantarum CJLP243	++	+	+	+	+	−
L. plantarum CJNR02	−	+	−	−	−	+
L. plantarum CJNR04	−	+	−	−	−	+
L. plantarum CJNR16	−	+	−	+	−	+
L. plantarum CJNR26	−	+	++	+	+	++
L. plantarum CJNR37	−	+	++	−	+	++
L. plantarum CJNR54	−	+	++	+	++	+
L. plantarum BJ53	+	−	+	+	−	−
L. plantarum BJ54	+	−	−	+	−	−
L. plantarum BJ57	−	−	+	−	−	−
L. plantarum BJ58	+	+	++	+	+	−
L. plantarum BJ135	+	+	−	+	+	−
L. plantarum BJ203	+	−	−	+	+	−
L. plantarum BJ217	+	+	+	+	+	−
L. plantarum BJ224	+	−	+	−	++	+
L. plantarum BJ242	+	+	++	+	+	−

Figure 71. Inhibition of six intestinal pathogens by 59 lactobacilli [194].

Ringel-Kulka et al., (2011) investigated the clinical efficacy of *Lactobacillus acidophilus* NCFM (L-NCFM) and *Bifidobacterium lactis* Bi-07 (B-LBi07) in non-constipation FBDs [195]. A double-blind, placebo-control clinical trial of the probiotic bacterias L-NCFM and B-LBi07 was conducted twice a day (2 x 10^{11}CFU/d) versus placebo over 8 weeks. Primary endpoints were global relief of gastrointestinal symptoms and satisfaction with treatment. Secondary endpoints were change in symptoms severity, well-being, and quality of life. Microbiological effect was assessed by quantitative real time PCR on fecal samples. Sixty patients (probiotic, n = 31; placebo, n = 29), 72% females, 84% whites, mean age 37 years. Abdominal bloating improved in the probiotics compared with the placebo group at 4 weeks (4.10 vs 6.17, p = 0.009; change in bloating severity p = 0.02) and 8 weeks (4.26 vs 5.84, p = 0.06; change in bloating severity $p < 0.01$). Analyses on the irritable bowel syndrome subgroup (n = 33) showed similar results. L-NCFM and B-LBi07 twice a day improve symptoms of bloating in patients with FBDs. These data supported the role of intestinal bacteria in the pathophysiology of FBD and the role for probiotic bacteria in the management of these disorders.

Simakachorn et al., (2011) demonstrated the tolerance and safety of an enteral formula containing prebiotics/probiotics, and its effect on the faecal microbiota in critically ill children. Ninety-four patients between 1 and 3 years old under mechanical ventilation requiring enteral feeding were randomized to receive either a test formula containing a synbiotic blend (*Lactobacillus paracasei* NCC 2461 and *Bifidobacterium longum* NCC 3001), fructooligosaccharides, inulin, and Acacia gum), or a control formula [196]. Patients remained in the intensive care unit for 7 days and were examined at day 14. Tolerance was assessed by overall caloric intake and time to reach caloric goal. Safety was assessed by abdominal distention, vomiting, and stool frequency. Microbiota was analyzed by culture- and molecular-based methods. Overall caloric intake and time to reach caloric goal were similar between groups. Abdominal distention, vomiting, and stool frequency were not affected by the supplementation with pre- and probiotics. Faecal Bifidobacteria were higher in the test group at the end of the study. A similar trend was observed for total lactobacilli. *Lactobacillus paracasei* NCC 2461 and *Bifidobacterium longum* NCC 3001 were detected in 80.4% and 17% of the test group patients, respectively. Enterobacteria levels remained unchanged during hospitalization in the control group but diminished in the test group. The enteral formula supplemented with synbiotics was well tolerated by children in intensive care units; it was safe and produced an increase in faecal bacterial groups of previously reported beneficial effects.

Hütt et al., (2011) evaluated the safety and persistence of selected *Lactobacillus* strains in the gastrointestinal tract of healthy adult volunteers after oral consumption of high doses of lactobacilli to identify potential candidates for probiotic and biotechnological applications [197]. In the first phase of the study, nine individuals consumed capsules containing

Lactobacillus gasseri 177 and E16B7, *Lactobacillus acidophilus* 821-3, *Lactobacillus paracasei* 317 and *Lactobacillus fermentum* 338-1-1 (each daily dose 1×10^{10} CFU) for 5 consecutive days. The persistence of *Lactobacillus* strains was assessed by culturing combined with arbitrarily primed polymerase chain reaction (AP-PCR) and PCR-Denaturing Gradient Gel Electrophoresis (PCR-DGGE) on days 0, 5, 8, 10 and 20 from faecal samples. All strains survived gastrointestinal passage and were detected on the 5th day. *Lactobacillus acidophilus* 821-3 was detected in four volunteers on the 8th day (4.3 to 7.0 Log CFU/g) and in two on the 10th day (8.3 and 3.9 Log CFU/g, respectively). In the second phase of the study, five additional volunteers consumed *Lactobacillus acidophilus* 821-3 (daily 1×10^{10} CFU) for 5 consecutive days. The strain was subsequently detected in faeces of all individuals using real-time PCR on the 10th day (range 4.6-6.7; median 6.0 Log cell/g) in both phases of the study for at least 5 days after discontinuation of consumption. The administration of high doses of different *Lactobacillus* strains did not result in any severe adverse effects in GIT and/or abnormal values of blood indices. The strain *Lactobacillus acidophilus* 821-3 could be a promising candidate for probiotic and biotechnological applications.

Klein (2011) investigated 92 strains from the genus *Lactobacillus* isolated from probiotics, food, and clinical sources [198]. Species tested were the *Lactobacillus acidophilus group, Lactobacillus casei* group, *Lactobacillus reuteri/fermentum group,* and *Lactobacillus sakei/curvatus* group. Cell and colony morphology, fermentation patterns, and growth characteristics as well as soluble whole cell proteins were analyzed. Antibiotic resistance against clinically important agents was determined by broth dilution tests. The vanA and tet genes were confirmed. Resistances occurred mainly against gentamicin, ciprofloxacin, clindamycin, sulfonamides, and, in some cases, glycopeptides. The natural glycopeptide resistance within the *Lactobacillus casei* group and *Lactobacillus reuteri* appears to be not of clinical relevance, as there was no vanA gene present. Therefore, the transfer of this resistance is very unlikely. Tet-(A), -(B), -(C), -(M), or -(O) gene could not be detected. The protein fingerprinting within the *Lactobacillus casei* group proved that *Lactobacillus rhamnosus* strains of clinical origin clustered together with probiotic strains. For safety evaluations resistance patterns of a broad range of strains are a useful criterion together with the exclusion of known resistance genes (like the vanA gene) and can be used for decision making on the safety of probiotics, both by authorization bodies and manufacturers.

Schiffer et al., (2011) studied *in vivo* the effects of one strain of *Lactobacillus casei* on IgE-induced passive systemic anaphylaxis and IgG-induced passive arthritis, two murine models of acute allergic and autoimmune inflammation, respectively, which bypass the induction phase of immune responses, *in vitro* on IgE- and IgG-induced mouse mast cell activation and *ex vivo* on IgE-dependent human basophil activation [199]. *Lactobacillus casei* protected from anaphylaxis and arthritis, and inhibited mouse mast cell and human basophil activation. Inhibition required contact between mast cells and bacteria, was reversible, and selectively affected the Lyn/Syk/linker for activation of T cells pathway induced on engagement of IgE receptors, leading to decreased MAPK activation, Ca_2^+ mobilization, degranulation, and cytokine secretion. Also, adoptive anaphylaxis induced on Ag challenge in mice injected with IgE-sensitized mast cells was abrogated in mice injected with IgE-sensitized mast cells exposed to bacteria. These results demonstrate that probiotics can influence the effector phase of adaptive immunity in allergic and autoimmune diseases. They might, therefore, prevent inflammation in patients who have already synthesized specific IgE or autoantibodies.

Shimosato et al., (2011) conducted an experiment with the objective of enhancing mucosal immunity by administration of *Lactobacillus pentosus* ONRIC b0240 (b240) in PEM model mice [200]. Three groups of male C3H/HeN mice aged approximately 12 weeks were caged in groups of five or six and received various treatments. Mice were fed 4 (low-protein diet; PEM model), 20 (standard-protein diet), or 40% (high-protein diet) ovalbumin (OVA) with or without 0.05% b240. Five weeks later, all mice were sacrificed, and the organs were extracted for analysis of the immune response. Acute toxicity was not observed and the addition of b240 showed no influence on body weight; however, body weight decreased with increasing protein level. Interestingly, intestinal total IgA was significantly increased ($p < 0.05$) in all test diets with b240. The *in vitro* study showed that the number of B cells and type 2 helper T (T_H2) cells were significantly increased in mouse spleen cells with b240 treatment, whereas no differences were found in the number of T_H1 cells. b240 also had the ability to augment IgA and IgG production in mouse Peyer's patch cells. Thus b240 enhanced IgA production and helped recover the intestinal immune system in PEM model mice via augmentation of humoral immunity.

Mangell et al., (2012) investigated the effect of the probiotic *Lactobacillus plantarum* 299v on the intestinal load of potentially pathogenic bacteria, bacterial translocation, and cell proliferation in elective colon surgery [201]. Seventy-five patients were randomized to pre- and postoperative oral intake of *Lactobacillus plantarum* 299v or placebo. Rectal swabs and mucosal biopsies were taken before the start of intake, after one week, at surgery, and after six days, weeks, and months. Viable counts were quantified for Clostridia, Enterobacteriaceae, Gram-negative anaerobes, and Lactobacilli. Bacterial translocation was determined by the analysis of bacterial DNA genes in mesenteric lymph nodes. Ki-67 was used as a marker of cell proliferation in normal mucosa and tumor. *Lactobacillus plantarum* 299v was given without adverse effects. *Lactobacillus plantarum* 299v as well as *Enterobacteriaceae* and Gram-negative anaerobes increased in the colon one week after the administration of *Lactobacillus plantarum* 299v. There were no significant differences between patients receiving *Lactobacillus plantarum* 299v and placebo in the incidence of bacterial translocation (27 vs. 13 %) and postoperative complications (16 vs. 31 %). *Lactobacillus plantarum* 299v was established in the intestine, but no inhibitory effect on enteric bacteria, bacterial translocation, or postoperative complications was found. The mechanism behind the protective effects of probiotics found in animal and some human studies remain elusive and require further explorations. No adverse effects were recorded after the administration of high doses of *Lactobacillus plantarum* 299v.

Holcher et al., (2012) aimed to investigate the effect of infant starter formula containing the probiotic *Bifidobacterium animal*is subsp. *lactis* (Bb12) on intestinal immunity and inflammation [202]. Six-week-old healthy, full-term infants (n = 172) were enrolled in a prospective, randomized, double-blind, controlled clinical trial with 2 groups studied in parallel to a breastfed comparison group. Formula-fed (FF) infants were randomized to partially hydrolyzed whey formula (CON) or the same formula containing 10^6 CFU Bb12/g (PRO) for 6 weeks. Fecal secretory IgA (sIgA), calprotectin, lactate, and stool pH were assessed at baseline, 2 weeks, and 6 weeks. Anti-poliovirus-specific IgA and anti-rotavirus-specific IgA were assessed at 2 and 6 weeks. Among vaginally delivered FF infants, PRO consumption increased ($p < 0.05$) fecal sIgA compared to CON. Anti-poliovirus-specific IgA concentration increased ($p < 0.05$) in all infants consuming PRO, whereas anti-rotavirus-specific IgA tended to increase ($p = 0.056$) with PRO consumption in cesarean-delivered infants. Anthropometrics and tolerance did not differ significantly between FF infants. Infants consuming formula with Bb12 produced feces with detectable presence of Bb12 and augmented sIgA concentration. Furthermore, cesarean-delivered infants consuming Bb12 had heightened immune response, as evidenced by increased anti-rotavirus- and anti-poliovirus-specific IgA following immunization. Negative immune-related effects of not breastfeeding and cesarean delivery could be mitigated by including Bb12 in infant formula, thereby providing infants a safe, dietary, immune-modulating bacterial introduction.

Gouriet et al., (2012) conducted a retrospective study, which consisted of the screening of *Lactobacillus* isolates obtained in their laboratory from January 2004 to December 2009 [203]. The polymerase chain reaction (PCR) assay was selected as the gold standard method. The isolates were first identified using API Coryne strips, matrix-assisted laser desorption/ionization time-of-flight mass spectrometry (MALDI-TOF MS), and 16S rRNA gene sequencing. *Lactobacillus* tuf gene-based identification was used when the 16S rRNA results were inconclusive. Among the 60 strains of *Lactobacillus* spp., *Lactobacillus rhamnosus* was the most commonly isolated species and was found in blood cultures from 16 patients. Combined with 45 patients reported in the literature, patients presenting with *Lactobacillus rhamnosus* bacteremia experienced nosocomial infections associated with both immunosuppression (66 %) and catheters (83 %).

Maccaferri et al., (2012) have recently demonstrated that *Kluyveromyces marxianus* B0399 is highly adhesive to human enterocyte-like CaCO-2 cells and that in the presence of inflammatory stimulation with lipopolysaccharide, it was able to provoke a decreases in the levels of production of proinflammatory cytokines in PBMCs and CaCO-2 cells, thus ameliorating the inflammatory response [204]. Moreover *Kluyveromyces marxianus* B0399 increased the bifidobacterial concentration in the stages of the colonic model system simulating the proximal and transverse colon flora. The amounts of the short-chain fatty acids acetate and propionate also increased following yeast supplementation. Finally, *Kluyveromyces marxianus* B0399 induced a decrease of the cytotoxic potential of the culture supernatant from the first stage of the colonic model system.

Finamore et al., (2012) evaluated the modulation of intestinal and systemic antigen-specific immune response by *Lactobacillus rhamnosus* GG or *Bifidobacterium animalis* MB5 in tolerized and immunized rats [205]. Three groups of rats received orally *Lactobacillus rhamnosus* GG, *Bifidobacterium animalis* MB5, or PBS (control) for 28 days. Each group was divided into two subgroups of tolerized or immunized rats receiving orally ovalbumin (OVA; 7 mg) or PBS on day 7, 9, and 11. All rats were immunized with OVA (300 μg) on d 14 and 21. In tolerized rats, the OVA-induced proliferative response of mesenteric lymph nodes (MLN) and spleen cells did not differ from control, indicating that the two probiotics maintained the tolerance. *Lactobacillus rhamnosus* GG and *Bifidobacterium animalis* MB5 in immunized rats reduced the OVA-induced proliferative response in MLN ($p <$

0.01) but not in spleen, whereas the proliferative response to anti-CD3 and concanavalin A of MLN and spleen cells as well as the delayed-type hypersensitivity reaction were not affected by probiotic treatment, indicating OVA-specific hyporesponsiveness restricted to intestinal immunity. This hyporesponsiveness was associated with $CD4^{+}CD25^{+}Foxp3^{+}$ T cell expansion ($p < 0.01$) and increased IL-10 and TGFβ after *Lactobacillus rhamnosus* GG ($p < 0.05$), and increased apoptosis after *Bifidobacterium animalis* MB5 ($p < 0.001$) in MLN. In conclusion, a novel activity of LGG and *Bifidobacterium animalis* in inducing OVA-specific hyporesponsiveness in MLN of OVA-immunized rats was reported; such activity can be useful for therapeutic strategies aimed to prevent undesirable reactions to immunogenic antigens in the gut.

After several efforts to produce effective vaccines against Bovine herpesvirus 5 (BoHV-5), Roos et al., (2012) observed a systemic enhanced IgG antibody response toward Th1, and an increased IFN-γ, IL-12 and IL-10 cytokines mRNA levels in animals supplemented with the probiotic *Bacillus cereus* var. *toyoi* [206], Such results resulted a promising means of improving vaccines efficacy, in particular those which rely on a cell-mediated immune response.

Fanning et al., (2012) demonstrated that *Bifidobacterium breve* UCC2003 was able to produce a cell surface-associated exopolysaccharide (EPS) which provided stress tolerance and promoted *in vivo* persistence, but not initial colonization [207]. EPS production was shown to be linked to the evasion of adaptive B-cell responses and EPS(+) *Bifidobacterium breve* UCC2003 reduced colonization levels of the gut pathogen *Citrobacter rodentium*, thus providing pathogen protection.

The mechanism of TNF inhibition by *Lactobacillus reuteri* 6475-derived histamine, investigated by Thomas et al., (2012) using Toll-like receptor 2 (TLR2)-activated human monocytoid cells, showed that histamine stimulated increased levels of cAMP, which inhibited downstream MEK/ERK MAPK signaling via protein kinase A (PKA) and resulted in suppression of TNF production by transcriptional regulation [208].

In the study of Platinga et al., (2012), cytokine production induced by PRR ligands or *Candida albicans* was assessed in conditions of simultaneous stimulation or preincubation of primary immune cells with three different potentially probiotic strains, *Bifidobacterium breve* (NumRes 204), *Lactobacillus rhamnosus* (NumRes1) and *Lactobacillus casei* (DN-114 001) [209]. Simultaneous stimulation led to a potentiation of IL-1β and IL-6 production, while the TNFα and IFN-γ production was inhibited. After pre-incubation with these potentially probiotic a lower production of TNFα was observed in the presence of *Bifidobacterium breve* (NumRes 204). Moreover, *Candida albicans*-induced IL-17 production was decreased after pre-incubation with both *Bifidobacterium* or *Lactobacillus* probiotic strains.

Zhong et al., (2012) investigated the effect of living probiotics, probiotic DNA and the synthetic oligodeoxynucleotides containing CpG motifs (CpG-ODN) on both immune response and intestinal barrier function in ovalbumin-sensitized rat and the underlying mechanisms [210].

Brown-Norway rats were orally sensitized with ovalbumin, and living probiotics, probiotic DNA extraction, synthetic CpG-ODN or non-CpG ODN control was administered. In the living probiotics, probiotic DNA and CpG-ODN groups, the allergic response was significantly inhibited, the Th1/Th2 cytokine balance was shifted away from Th2 side, the percentage of $CD4(^{+})$ $CD25(^{+}high)$ Treg cells was increased, and the intestinal barrier function was improved. The levels of toll-like receptor (TLR) 9 mRNA and nuclear factor (NF)-κB activity, as well as the IκB-α phosphorylation (p-IκB-α) was significantly increased in these three intervention groups compared with the OVA-positive group, whereas no such effects were found in the non-CpG ODN control group.

In summary, probiotic genomic DNA and the synthetic CpG-ODN was comparable with living probiotics in preventing food allergic response by immune modulation and intestinal barrier function enhancement, and the activation of TLR9/NF-κB signal pathway might be involved in this process. Oral administration of *Lactobacillus paracasei* NCC 2461 was shown to protect mothers against the development of airway inflammation in offspring by activating regulatory pathways, likely through TLR2/4 signalling [211].

More in detail, attenuation of eosinophil influx in the lungs; reduction of IL-5 levels in bronchoalveolar lavage, and in lung and mediastinal lymph node cell cultures; and reduced peribronchial inflammatory infiltrate and mucus hypersecretion were observed. While allergen-specific IgE and IgG antibody levels remained unchanged by the treatment, IL-4 and IL-5 production in spleen cell cultures were significantly reduced upon allergen stimulation in offspring of *Lactobacillus paracasei* treated mice. Offspring of *Lactobacillus paracasei* supplemented mothers had significantly reduced Bet v 1-specific as well as Concanavalin A-induced responses in spleen and mesenteric lymph node cell cultures, suggesting the modulation of both antigen-specific and mitogen-induced immune responses in offspring. These effects were associated with increased Foxp3 mRNA expression in the lungs and increased

TGF-beta in serum, based upon the fact that liver X receptors (LXRs) play a role in cholesterol efflux.

Dong et al., (2012) compared the immunomodulatory properties of six probiotic strains of different species and two genera in a human peripheral blood mononuclear cell (PBMC) model *in vitro* [212]. Live cells of lactobacilli (*Lactobacillus casei* Shirota, *Lactobacillus rhamnosus* GG, *Lactobacillus plantarum* NCIMB 8826 and *Lactobacillus reuteri* NCIMB 11951) and bifidobacteria (*Bifidobacterium longum* SP 07/3 and *Bifidobacterium bifidum* MF 20/5) were individually incubated with PBMC from seven healthy subjects for 24 hours.

Probiotic strains increased the proportion of $CD69^+$ on lymphocytes, T cells, T cell subsets and natural killer (NK) cells, and increased the proportion of $CD25^+$, mainly on lymphocytes and NK cells. The effects on activation marker expression did not appear to be strain specific (figure 72 A-B).

NK cell activity was significantly increased by all six strains, without any significant difference between strains. Probiotic strains increased production of IL-1β, IL-6, IL-10, TNF-α, granulocyte-macrophage colony-stimulating factor and macrophage inflammatory protein 1α to different extents, but had no effect on the production of IL-2, IL-4, IL-5 or TNF-β. The cytokines that showed strain-specific modulation included IL-10, INF-γ, TNF-α, IL-12p70, IL-6 and monocyte chemotactic protein-1. The *Lactobacillus* strains tended to promote T helper 1 cytokines, whereas bifidobacterial strains tended to produce a more anti-inflammatory profile. There was limited evidence of strain-specific effects of probiotics with respect to T cell and NK cell activation or NK cell activity, whereas production of some cytokines was differentially influenced by probiotic strains.

Karlsson et al., (2012), tested twelve *Lactobacillus* strains for their ability to influence CXCL8 release from urothelial cells [213]. CXCL8 is an inflammatory chemokine released by the urinary tract mucosa after contact with uropathogenic *Escherichia coli* during urinary tract infection and is crucial for proper infiltration of immune cells.

A majority of the tested species altered CXCL8 release from the urothelial cells after 24 hours of stimulation. Most species increased CXCL8 release, whereas a few lactobacilli efficiently suppressed CXCL8 secretion from *Escherichia coli*-challenged cells. While strong CXCL8 modulators such as *Lactobacillus reuteri* and *Lactobacillus delbrueckii* were unable to degrade CXCL8 in the extracellular environment, effects on IL8 transcription were evident for selected lactobacilli. Although IL8 transcription was affected by

(A)

Treatment	CD3⁺							
	Positive (%)				MFI			
	CD69		CD25		CD69		CD25	
	Mean	SE	Mean	SE	Mean	SE	Mean	SE
Medium	0·71	0·28	3·39	0·27	24·39	1·33	23·92	1·05
L. casei Shirota	5·36*	1·52	5·22	0·25	43·27*	3·76	26·02	0·49
L. rhamnosus GG	3·67	1·17	3·59	0·29	37·17	3·28	22·54	0·63
L. plantarum NCIMB 8826	5·34*	1·57	6·12	0·66	48·43***	4·54	32·08***	2·06
L. reuteri NCIMB 11951	5·49*	2·00	5·22	1·00	52·10***	6·70	25·60	1·35
B. longum SP 07/3	4·13	1·17	5·15	0·31	43·65*	3·05	26·96	0·64
B. bifidum MF 20/5	3·41	1·19	5·13	0·30	36·93	2·08	26·50	0·63

Treatment	$CD3^+CD4^+$							
	Positive (%)				MFI			
	CD69		CD25		CD69		CD25	
	Mean	SE	Mean	SE	Mean	SE	Mean	SE
Medium	0·50	0·19	7·10	0·81	18·29	2·94	17·08	1·07
L. casei Shirota	2·47	0·72	9·99	0·77	23·76	1·06	19·62	0·67
L. rhamnosus GG	1·72	0·45	8·01	0·71	22·41	1·82	17·20	0·61
L. plantarum NCIMB 8826	2·59	0·75	11·18	0·66	24·44	1·12	21·72*	0·97
L. reuteri NCIMB 11951	3·18*	0·95	10·37	0·90	24·04	0·96	19·85	1·19
B. longum SP 07/3	1·52	0·36	10·16	1·12	24·21	1·61	19·42	0·79
B. bifidum MF 20/5	1·37	0·48	9·54	1·00	25·64	2·72	20·50	0·61

Treatment	$CD3^+CD8^+$							
	Positive (%)				MFI			
	CD69		CD25		CD69		CD25	
	Mean	SE	Mean	SE	Mean	SE	Mean	SE
Medium	1·58	0·59	0·99	0·49	24·61	1·21	22·84	1·26
L. casei Shirota	11·49***	2·48	2·47	0·76	46·41*	5·74	24·30	0·71
L. rhamnosus GG	7·63**	1·82	1·75	0·43	38·45	3·69	23·88	1·52
L. plantarum NCIMB 8826	11·27***	2·36	3·53	1·05	51·00**	5·49	31·94*	1·71
L. reuteri NCIMB 11951	10·87***	3·59	4·16	1·27	49·45*	5·83	25·86	1·99
B. longum SP 07/3	8·86***	1·81	2·22	0·49	46·73*	4·15	25·6	1·07
B. bifidum MF 20/5	7·65**	2·43	1·98	0·59	39·05	3·23	26·62	1·27

(B)

Treatment	$CD56^+CD3^-$							
	Positive (%)				MFI			
	CD69		CD25		CD69		CD25	
	Mean	SE	Mean	SE	Mean	SE	Mean	SE
Medium	3·80	0·56	0·18	0·04	27·00	1·38	18·98	1·95
L. casei Shirota	78·68***	3·87	21·18***	1·67	60·27**	5·71	24·28	0·52
L. rhamnosus GG	61·42**	6·16	8·08*	1·09	41·15	1·63	21·10	0·51
L. plantarum NCIMB 8826	79·95***	3·24	20·58***	2·63	67·55***	6·69	23·52	0·55
L. reuteri NCIMB 11951	82·52***	2·85	19·40**	4·47	69·53***	9·87	21·86	1·33
B. longum SP 07/3	75·93***	4·17	22·26***	2·85	59·17*	6·89	25·54*	0·72
B. bifidum MF 20/5	57·55**	5·81	11·49*	2·76	42·47	4·41	25·06*	0·72

Figure 72. (A) The activation of T lymphocyte subsets by different probiotic strains. MFI, mean fluorescence intensity; L., *Lactobacillus*; B., *Bifidobacterium*; (B) Activation of natural killer cells by different probiotic strains. MFI, mean fluorescence intensity; L., *Lactobacillus*; B., *Bifidobacterium* [212].

lactobacilli, the influence on mRNA transcript did not correlate to the impact on CXCL8 release.

Lactobacillus salivarius SMXD51, which produces a bacteriocin-like substance highly active against the foodborne pathogen *Campylobacter jejuni*, resluted to be tolerant to gastrointestinal conditions (acid and bile), to adhere to intestinal cells and stimulate the immune system [214]. The bacterium strengthened the intestinal barrier functions through the increase of the transepithelial electrical resistance (TEER) and reinforcement of the F-actin cytoskeleton. One hour pretreatment with *Lactobacillus salivarius* SMXD51 protected against *Pseudomonas aeruginosa* PAO1-induced decrease of TEER and damage of the F-actin cytoskeleton. Gut immune system is daily exposed to a plethora of antigens contained in the environment as well as in food. Both secondary lymphoid tissue, such as Peyer's patches, and lymphoid follicles (tertiary lymphoid tissue) are able to respond to antigenic stimuli releasing cytokines or producing antibodies (secretory IgA) [215].

Intestinal epithelial cells are in close cooperation with intraepithelial lymphocytes and possess Toll-like receptors on their surface and Nod-like receptors (NLRs) which sense pathogens or pathogen-associated molecular patterns. Intestinal microbiota, mainly composed of *Bacteroides* and *Firmicutes*, generates tolerogenic response acting on gut dendritic cells and inhibiting the T helper (h)-17 cell anti-inflammatory pathway. This is the case of *Bacteroides fragilis* which leads to the production of interleukin-10, an anti-inflammatory cytokine, from both T regulatory cells and lamina propria macrophages. Conversely, segmented filamentous bacteria rather induce Th17 cells, thus promoting intestinal inflammation. Intestinal microbiota and its toxic components have been shown to act on both Nod1 and Nod2 receptors and their defective signaling accounts for the development of inflammatory bowel disease (IBD). In IBD a loss of normal tolerance to intestinal microbiota seems to be the main trigger of mucosal damage. In addition, intestinal microbiota thanks to its regulatory function of gut immune response can prevent or retard neoplastic growth. In fact, chronic exposure to environmental microorganisms seems to be associated with low frequency of cancer risk. Major nutraceuticals or functional foods employed in the modulation of intestinal microbiota are represented by prebiotics, probiotics, polyunsaturated fatty acids, amino acids and polyphenols. The cellular and molecular effects performed by these natural products in terms of modulation of the intestinal microbiota and mostly attenuation of the inflammatory pathway are described.

Yoon et al., (2013) confirmed the effects of *Lactobacillus rhamnosus* BFE5264, *Lactobacillus plantarum* NR74 and *Lactobacillus rhamnosus* GG on cholesterol efflux, by upregulatig ABCA1 and ABCG1 expression, and anti-inflammatory action in macrophages, by suppressing expression of interleukin (IL)-1β and tumor necrosis factor (TNF)-α [216].

Taverniti et al. (2013) characterized *Lactobacillus helveticus* MIMLh5 and its surface-layer protein (SlpA) which were found to exert anti-inflammatory effects by reducing the activation of NF-κB on the intestinal epithelial CaCO-2 cell line [217]. On the contrary, MIMLh5 and SlpA can act as stimulators of the innate immune system by triggering the expression of proinflammatory factors tumor necrosis factor alpha and COX-2 in the human macrophage cell line U937 via recognition through Toll-like receptor 2. In the same experiments, SlpA protein did not affect the expression of the anti-inflammatory cytokine interleukin-10. A similar response was observed following stimulation of macrophages isolated from mouse bone marrow or the peritoneal cavity.

Carey et al., (2013) investigated modulation of the inflammatory host response induced by *Salmonella enterica* serovar *typhimurium* DT104 in the presence of selected probiotics and LAB isolated from human sources, dairy products, and farm animals. IL-8 gene expression and protein production in HT-29 cells were evaluated by real-time PCR and ELISA, respectively [218]. Pre-incubation of HT-29 cells with *Lactobacillus kefir* IM002, *Bifidobacterium adolescentis* FRP 61, *Bifidobacterium longum* FRP 68 and FRP 69, *Bifidobacterium breve* FRP 334, and *Leuconostoc mesenteroides* IM080 significantly inhibited IL-8 secretion induced by *Salmonella typhimurium* DT104. Co-culture of selected probiotics and *Salmonella typhimurium* DT104 reduced IL-8 production, while potential probiotics and LAB had no effect on IL-8 secretion in HT-29 cells preincubated with *Salmonella typhimurium* DT104 prior to adding probiotics. *Lactobacillus kefir* IM002 supernatant also significantly reduced IL-8 production. In conclusion, the study suggested that probiotic bifidobacteria and LAB modulate cytokine induction and possess anti-inflammatory properties; however, the effectiveness is strain dependent.

Cytokine responses in the Japanese pufferfish (Takifugu rubripes) head kidney (HK) cells to heat-killed probiotic LAB isolated from the Mongolian dairy products were investigated by Biswas et al., (2013) [219]. HK cells were incubated with two heat-killed bacteria, namely *Lactobacillus paracasei* spp. *paracasei* (strain 06TCa22) and *Lactobacillus plantarum* (strain 06CC2) and the responses of 16 cytokine genes at 0 (control), 1, 4, 8, 12, 24 and 48 hours post-stimulation were assayed by multiplex RT-PCR analysis. Despite the differences in the transcriptional profiles, expression of all the cytokines tested here was significantly elevated by both the probiotic bacterial stimulants compared with the unstimulated control. *Lactobacillus reuteri* DSM 17938 improved survival and reduced the incidence and severity of NEC in a rodent model [220]. Foxp3(+) regulatory T cells (Tregs) maintain intestinal homeostasis by controlling inflammation and inducing tolerance. To determine whether there are insufficient numbers of Tregs to control inflammation in NEC and to determine if LR17938 increases the frequency of Tregs, studied selected groups of newborn Sprague-Dawley rats according to feeding plan: dam ± LR17938, formula ± LR17938, and NEC ± LR17938. NEC was induced by gavage feeding with special formula and exposure to hypoxic conditions. Lymphocytes isolated from ileum, mesenteric lymph nodes (MLN), spleen and thymus were labeled for T cell surface markers (CD3, CD4, CD8) and intracellular Foxp3; and labeled cells were analyzed by flow cytometry. The percentage of CD3(+) T cells and Foxp3(+) Tregs in the ileum significantly decreased in pups with NEC, compared to normal controls. Feeding LR17938 to neonatal rats with NEC increased the % of Foxp3(+) T cells in the ileum while decreasing the percentage of cells in the MLN. Administration of LR17938 to dam-fed rats significantly increased Foxp3(+)Tregs in the ileum as early as day of life (DOL)1 but did not produce an increase in Tregs in formula-fed rats on DOL1. These results suggest that factors in breast milk may enhance the early immunomodulatory effects of LR17938. An anti-inflammatory effect of LR17938 in NEC was associated with the modulation of immune responses and induction and what appears to be migration of Foxp3(+) Tregs to the diseased gut. Probiotic-facilitated development of Tregs might play an important role in the prevention of NEC.

Viable and heat-killed *Lactobacillus acidophilus* and *Lactobacillus paracasei*, which have properties of acid and bile tolerance and inhibitor effects on pathogens, were shown to induce low levels of proinflammatory cytokines (TNF-α, IL-6 and IL-8) and high levels of IFN-γ and IL-12p70 in PBMCs [221]. In anti-CD3 antibody activated PBMCs, viable and heat-killed C44 increased Th2 cytokine levels (IL-5, IL-6 and IL-10), and simultaneously enhanced Th1 responses by inducing IFN-γ and IL-12p70 production. Different from that of *Lactobacillus* strains, their genomic DNA induced low levels of IL-12p70, IFN-γ and proinflammatory cytokines in PBMCs with or without anti-CD3 antibody activation. These results provided *in vitro* evidence that the genomic DNA of strains of C44 and C48, especially C44, induced weaker inflammation, and may be potentially applied for treating allergic diseases.

Chapter V

Conclusion

Our review reports an exhaustive overview of the major experimental and clinical studies involving lactobacilli as potential immune-modulating agents, in order to prevent and/ or to treat, not only inflammatory bowel diseases, and related syndromes, but also to extend their potential use in the prophylaxis of cancer enhancing the immunosurveillance and in the control of the tumor growth by direct inhibition, and cytokine production. As stated by Taverniti et al., (2011), the use of both non-viable microbial cells and/or cell components has been shown to influence host's immune system [222] and therefore the future trends of probiotics administration seem to be projected to positively interfere with the gut immunocompetent structures, not only addressed to local disease control, but to expand also in the challenge of systemic immunological imbalances preventing further infectious or inflammatory diseases in other districts, such as skin, mucosae, lungs, spleen, lymph nodes and liver (figure 73) [223].

Nevetheless, as stated by Marco et al., (2013), health claims of probiotics are constrained by the limited number of quality interventions with a clear end-point consumer benefit and by the need for an increased understanding of mechanisms underlying their effects (figure 74) [224, 225].

Therefore, in this paper we have also reviewed the potential diffusion of probiotics, through the gut wall to the blood stream, and the risk of septicemia. Fortunately the evidence of septic complication is very rare, and almost related to intestinal mucosa impairment with altered permeability of the basal membrane and transmural diffusion of probiotics: their clinical advantages administration largely overwhelms the risk of complications, and antibiotic treatment can properly control such untoward effect.

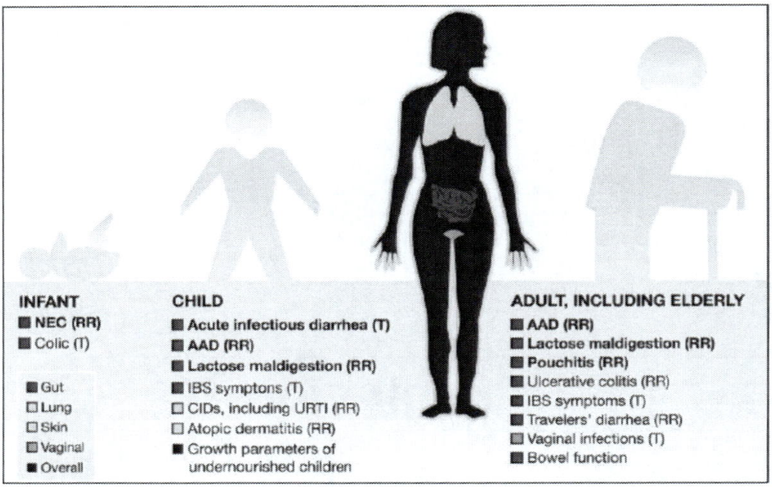

Figure 73. A schematic representation of manifold clinical actions of probiotics in some immunomediated pathologies (i.e., lactose maldigestion, atopic dermatitits, vaginal infections, etc) in humans [223].

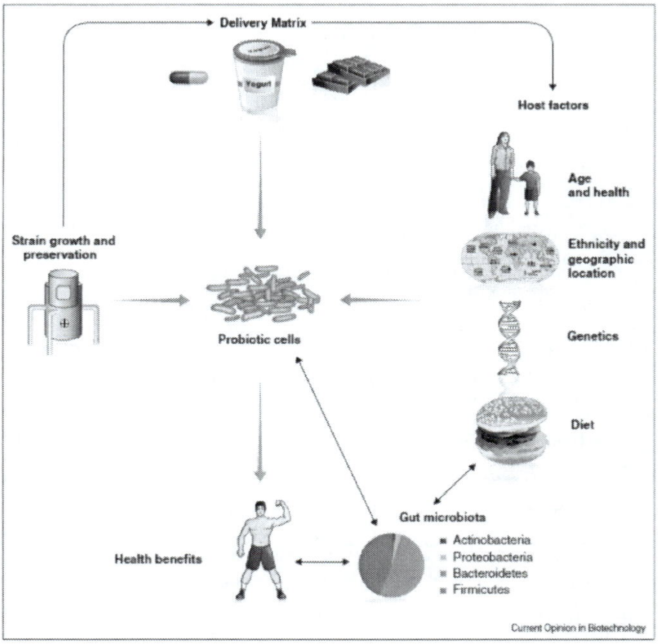

Figure 74. Schematic rapresentation of environmental determinants of probiotic efficacy [224].

This emerging issue may be acribed to the extensive use of live microbial cells [226] which has enhanced the interest in non-viable microorganisms or microbial cell extracts, as they could reduce shelf-life problems thus eliminating the risks of microbial translocation and infection for the consumers [227]. Based on this assumption taverniti introduced the term 'paraprobiotic' (or 'ghost probiotics'), to define either 'non-viable microbial cells (intact or broken) or crude cell extracts (i.e., with complex chemical composition) [222]

Chapter VI

References

[1] Saad N, Delattre C, Urdaci M, Schmitter JM, Bressollier P. An overview of the last advances in probiotic and prebiotic field. LWT - *Food Science and Technology*. 2013; 50(1):1–16.

[2] Elson G, Dunn-Siegrist I, Daubeuf B, Pugin J. Contribution of Toll-like receptors to the innate immune response to Gram-negative and Gram-positive bacteria. *Blood*. 2007; 109(4):1574-83.

[3] Aderem A, Ulevitch RJ. Toll-like receptors in the induction of the innate immune response. *Nature*. 2000; 406(6797):782-7.

[4] Beutler B, Hoebe K, Du X, Ulevitch RJ. How we detect microbes and respond to them: the Toll-like receptors and their transducers. *Journal of Leukocyte Biology*. 2003; 74(4):479-85.

[5] Kopp E, Medzhitov R. Recognition of microbial infection by Toll-like receptors. *Current Opinion in Immunology*. 2003; 15(4):396-401.

[6] Netea MG, van der Graaf C, Van der Meer JW, Kullberg BJ. Toll-like receptors and the host defense against microbial pathogens: bringing specificity to the innate-immune system. *Journal of Leukocyte Biology*. 2004; 75(5):749-55.

[7] Inohara, Chamaillard, McDonald C, Nunez G. NOD-LRR proteins: role in host-microbial interactions and inflammatory disease. *Annual Review of Biochemistry*. 2005; 74:355-83.

[8] Pugin J, Heumann ID, Tomasz A, Kravchenko VV, Akamatsu Y, Nishijima M, et al. CD14 is a pattern recognition receptor. *Immunity*. 1994; 1(6):509-16.

[9] Yasui H, Mike A, Ohwaki M. Immunogenicity of Bifidobacterium breve and change in antibody production in Peyer's patches after oral administration. *Journal of Dairy Science*. 1989; 72(1):30-5.

[10] Hockertz S. [Immunomodulating effect of killed, apathogenic Escherichia coli, strain Nissle 1917, on the macrophage system]. *Arzneimittel-Forschung*. 1991; 41(10):1108-12.

[11] Solis Pereyra B, Lemonnier D. Induction of 2'-5' A synthetase activity and interferon in humans by bacteria used in dairy products. *European Cytokine Network*. 1991; 2(2):137-40.

[12] Kaila M, Isolauri E, Soppi E, Virtanen E, Laine S, Arvilommi H. Enhancement of the circulating antibody secreting cell response in human diarrhea by a human Lactobacillus strain. *Pediatric research.* 1992; 32(2):141-4.

[13] Johansson ML, Molin G, Jeppsson B, Nobaek S, Ahrne S, Bengmark S. Administration of different Lactobacillus strains in fermented oatmeal soup: in vivo colonization of human intestinal mucosa and effect on the indigenous flora. *Applied and Environmental Microbiology*. 1993; 59(1):15-20.

[14] Perdigon G, Rachid M, De Budeguer MV, Valdez JC. Effect of yogurt feeding on the small and large intestine associated lymphoid cells in mice. *The Journal of Dairy Research*. 1994; 61(4):553-62.

[15] Link-Amster H, Rochat F, Saudan KY, Mignot O, Aeschlimann JM. Modulation of a specific humoral immune response and changes in intestinal flora mediated through fermented milk intake. *FEMS Immunology and Medical Microbiology*. 1994; 10(1):55-63.

[16] Aso Y, Akaza H, Kotake T, Tsukamoto T, Imai K, Naito S. Preventive effect of a Lactobacillus casei preparation on the recurrence of superficial bladder cancer in a double-blind trial. The BLP Study Group. *European Urology*. 1995; 27(2):104-9.

[17] Schiffrin EJ, Rochat F, Link-Amster H, Aeschlimann JM, Donnet-Hughes A. Immunomodulation of human blood cells following the ingestion of lactic acid bacteria. *Journal of Dairy Science*. 1995; 78(3):491-7.

[18] Mikes A, Ferencik M, Jahnova E, Ebringer L, Ciznar I. Hypocholesterolemic and immunostimulatory effects of orally applied Enterococcus faecium M-74 in man. *Folia microbiologica*. 1995; 40(6):639-46.

[19] Isolauri E, Joensuu J, Suomalainen H, Luomala M, Vesikari T. Improved immunogenicity of oral D x RRV reassortant rotavirus vaccine by Lactobacillus casei GG. *Vaccine.* 1995; 13(3):310-2.

[20] Majamaa H, Isolauri E, Saxelin M, Vesikari T. Lactic acid bacteria in the treatment of acute rotavirus gastroenteritis. *Journal of Pediatric Gastroenterology and Nutrition*. 1995; 20(3):333-8.

[21] Kaila M, Isolauri E, Saxelin M, Arvilommi H, Vesikari T. Viable versus inactivated Lactobacillus strain GG in acute rotavirus diarrhoea. *Archives of Disease in Childhood.* 1995; 72(1):51-3.

[22] Kalima P, Masterton RG, Roddie PH, Thomas AE. Lactobacillus rhamnosus infection in a child following bone marrow transplant. *The Journal of Infection.* 1996; 32(2):165-7.

[23] Sanyal D, Bhandari S. CAPD peritonitis caused by Lactobacillus rhamnosus. *The Journal of Hospital Infection. England* 1992. p. 325-7.

[24] Chomarat M, Espinouse D. Lactobacillus rhamnosus septicemia in patients with prolonged aplasia receiving ceftazidime-vancomycin. *European Journal of Clinical Microbiology and Infectious Disease.* 1991; 10(1):44.

[25] Rahman M. Chest infection caused by Lactobacillus casei ss rhamnosus. *British Medical Journal.* 1982; 284(6314):471-2.

[26] Griffiths JK, Daly JS, Dodge RA. Two cases of endocarditis due to Lactobacillus species: antimicrobial susceptibility, review, and discussion of therapy. *Clinical Infectious Diseases* 1992; 15(2):250-5.

[27] Naude WD, Swanepoel A, Bohmer RH, Bolding E. Endocarditis caused by Lactobacillus casei subspecies rhamnosus. A case report. *South African Medical Journal.* 1988; 73(10):612-4.

[28] Tenebaum MJ, Warner JF. Lactobacillus casei endocarditis. *Annals of Internal Medicine*. 1975; 82(4):539.

[29] Holliman RE, Bone GP. Vancomycin resistance of clinical isolates of lactobacilli. *The Journal of Infection.* 1988; 16(3):279-83.

[30] Golledge C. Vancomycin resistant lactobacilli. *The Journal of Hospital Infection.* 1988; 11(3):292.

[31] Miettinen M, Vuopio-Varkila J, Varkila K. Production of human tumor necrosis factor alpha, interleukin-6, and interleukin-10 is induced by lactic acid bacteria. *Infection and Immunity*. 1996; 64(12):5403-5.

[32] Cleveland MG, Gorham JD, Murphy TL, Tuomanen E, Murphy KM. Lipoteichoic acid preparations of Gram-positive bacteria induce interleukin-12 through a CD14-dependent pathway. *Infection and Immunity* 1996; 64(6):1906-12.

[33] Saxelin M, Chuang NH, Chassy B, Rautelin H, Makela PH, Salminen S, et al. Lactobacilli and bacteremia in southern Finland, 1989-1992. Clinical Infectious Diseases: an official publication of the *Infectious Diseases Society of America.* 1996; 22(3):564-6.

[34] Kishi A, Uno K, Matsubara Y, Okuda C, Kishida T. Effect of the oral administration of Lactobacillus brevis subsp. coagulans on interferon-alpha producing capacity in humans. *Journal of the American College of Nutrition.* 1996; 15(4):408-12.

[35] Larvol L, Monier A, Besnier P, Levecq H. Liver abscess caused by Lactobacillus acidophilus. *Gastroenterologie Clinique et Biologique.* 1996; 20(2):193-5.

[36] Malin M, Verronen P, Mykkanen H, Salminen S, Isolauri E. Increased bacterial urease activity in faeces in juvenile chronic arthritis: evidence of altered intestinal microflora? *British Journal of Rheumatology.* 1996; 35(7):689-94.

[37] Antony SJ, Stratton CW, Dummer JS. Lactobacillus bacteremia: description of the clinical course in adult patients without endocarditis. *Clinical Infectious Diseases.* 1996; 23(4):773-8.

[38] Sutas Y, Hurme M, Isolauri E. Down-regulation of anti-CD3 antibody-induced IL-4 production by bovine caseins hydrolysed with Lactobacillus GG-derived enzymes. *Scandinavian Journal of Immunology.* 1996; 43(6):687-9.

[39] Husni RN, Gordon SM, Washington JA, Longworth DL. Lactobacillus bacteremia and endocarditis: review of 45 cases. *Clinical Infectious Diseases.* 1997;25(5):1048-55.

[40] Majamaa H, Isolauri E. Probiotics: a novel approach in the management of food allergy. *Journal of Allergy and Clinical Immunology.* 1997; 99(2):179-85.

[41] Schiffrin EJ, Brassart D, Servin AL, Rochat F, Donnet-Hughes A. Immune modulation of blood leukocytes in humans by lactic acid bacteria: criteria for strain selection. *American Journal of Clinical Nutrition.* 1997; 66(2):515s-20s.

[42] Solis-Pereyra B, Aattouri N, Lemonnier D. Role of food in the stimulation of cytokine production. *American Journal of Clinical Nutrition.* 1997; 66(2):521s-5s.

[43] Marteau P, Vaerman JP, Dehennin JP, Bord S, Brassart D, Pochart P, et al. Effects of intrajejunal perfusion and chronic ingestion of Lactobacillus johnsonii strain La1 on serum concentrations and jejunal

secretions of immunoglobulins and serum proteins in healthy humans. *Gastroenterologie Clinique et Biologique*. 1997; 21(4):293-8.

[44] Sudo N, Sawamura S, Tanaka K, Aiba Y, Kubo C, Koga Y. The requirement of intestinal bacterial flora for the development of an IgE production system fully susceptible to oral tolerance induction. *Journal of Immunology*. 1997; 159(4):1739-45.

[45] Maassen CB, van Holten JC, Balk F, Heijne den Bak-Glashouwer MJ, Leer R, Laman JD, et al. Orally administered Lactobacillus strains differentially affect the direction and efficacy of the immune response. *The Veterinary Quarterly*. 1998; 20 Suppl 3:S81-3.

[46] Miettinen M, Matikainen S, Vuopio-Varkila J, Pirhonen J, Varkila K, Kurimoto M, et al. Lactobacilli and streptococci induce interleukin-12 (IL-12), IL-18, and gamma interferon production in human peripheral blood mononuclear cells. *Infection and Immunity*. 1998; 66(12):6058-62.

[47] Donnet-Hughes A, Rochat F, Serrant P, Aeschlimann JM, Schiffrin EJ. Modulation of nonspecific mechanisms of defense by lactic acid bacteria: effective dose. *Journal of Dairy Science*. 1999; 82(5):863-9.

[48] Spanhaak S, Havenaar R, Schaafsma G. The effect of consumption of milk fermented by Lactobacillus casei strain Shirota on the intestinal microflora and immune parameters in humans. *European Journal of Clinical Nutrition*. 1998; 52(12):899-907.

[49] Pelto L, Isolauri E, Lilius EM, Nuutila J, Salminen S. Probiotic bacteria down-regulate the milk-induced inflammatory response in milk-hypersensitive subjects but have an immunostimulatory effect in healthy subjects. *Clinical and Experimental Allergy*. 1998; 28(12):1474-9.

[50] Fukushima Y, Kawata Y, Hara H, Terada A, Mitsuoka T. Effect of a probiotic formula on intestinal immunoglobulin A production in healthy children. *International Journal of Food Microbiology*. 1998; 42(1-2):39-44.

[51] Rautio M, Jousimies-Somer H, Kauma H, Pietarinen I, Saxelin M, Tynkkynen S, et al. Liver abscess due to a Lactobacillus rhamnosus strain indistinguishable from L. rhamnosus strain GG. *Clinical Infectious Diseases*. 1999; 28(5):1159-60.

[52] Yoshimura A, Lien E, Ingalls RR, Tuomanen E, Dziarski R, Golenbock D. Cutting edge: recognition of Gram-positive bacterial cell wall components by the innate immune system occurs via Toll-like receptor 2. *Journal of Immunology*. 1999; 163(1):1-5.

[53] Madsen KL, Doyle JS, Jewell LD, Tavernini MM, Fedorak RN. Lactobacillus species prevents colitis in interleukin 10 gene-deficient mice. *Gastroenterology*. 1999; 116(5):1107-14.

[54] Perdigon G, Vintini E, Alvarez S, Medina M, Medici M. Study of the possible mechanisms involved in the mucosal immune system activation by lactic acid bacteria. *Journal of Dairy Science*. 1999; 82(6):1108-14.

[55] Herias MV, Hessle C, Telemo E, Midtvedt T, Hanson LA, Wold AE. Immunomodulatory effects of Lactobacillus plantarum colonizing the intestine of gnotobiotic rats. *Clinical and Experimental Immunology*. 1999; 116(2):283-90.

[56] Kato I, Tanaka K, Yokokura T. Lactic acid bacterium potently induces the production of interleukin-12 and interferon-gamma by mouse splenocytes. *International Journal of Immunopharmacology*. 1999; 21(2):121-31.

[57] Takagi A, Matsuzaki T, Sato M, Nomoto K, Morotomi M, Yokokura T. Inhibitory effect of oral administration of Lactobacillus casei on 3-methylcholanthrene-induced carcinogenesis in mice. *Medical Microbiology and Immunology*. 1999; 188(3):111-6.

[58] Miettinen M, Lehtonen A, Julkunen I, Matikainen S. Lactobacilli and Streptococci activate NF-kappa B and STAT signaling pathways in human macrophages. *Journal of Immunology*. 2000; 164(7):3733-40.

[59] Vitini E, Alvarez S, Medina M, Medici M, de Budeguer MV, Perdigon G. et al. Gut mucosal immunostimulation by lactic acid bacteria. *Biocell*. 2000; 24(3):223-32.

[60] Nagao F, Nakayama M, Muto T, Okumura K. Effects of a fermented milk drink containing Lactobacillus casei strain Shirota on the immune system in healthy human subjects. *Bioscience, Biotechnology, and biochemistry*. 2000; 64(12):2706-8.

[61] Hessle C, Andersson B, Wold AE. Gram-positive bacteria are potent inducers of monocytic interleukin-12 (IL-12) while Gram-negative bacteria preferentially stimulate IL-10 production. *Infection and Immunity*. 2000; 68(6):3581-6.

[62] Haller D, Bode C, Hammes WP, Pfeifer AM, Schiffrin EJ, Blum S. Non-pathogenic bacteria elicit a differential cytokine response by intestinal epithelial cell/leucocyte co-cultures. *Gut*. 2000; 47(1):79-87.

[63] Maassen CB, van Holten-Neelen C, Balk F, den Bak-Glashouwer MJ, Leer RJ, Laman JD, et al. Strain-dependent induction of cytokine profiles in the gut by orally administered Lactobacillus strains. *Vaccine*. 2000; 18(23):2613-23.

[64] Arunachalam K, Gill HS, Chandra RK. Enhancement of natural immune function by dietary consumption of Bifidobacterium lactis (HN019). *European Journal of Clinical Nutrition.* 2000; 54(3):263-7.

[65] Chiang BL, Sheih YH, Wang LH, Liao CK, Gill HS. Enhancing immunity by dietary consumption of a probiotic lactic acid bacterium (Bifidobacterium lactis HN019): optimization and definition of cellular immune responses. *European Journal of Clinical Nutrition.* 2000; 54(11):849-55.

[66] He FT, E.; Arvilommi, H.; Salminen, S. Modulation of humoral immune response through probiotic intake. *FEMS Immunlogy and Medical Microbiology.* 2000; 29:47-52.

[67] Isolauri E, Arvola T, Sutas Y, Moilanen E, Salminen S. Probiotics in the management of atopic eczema. *Clinical and Experimental Allergy: journal of the British Society for Allergy and Clinical Immunology.* 2000; 30(11):1604-10.

[68] Gill HS, Rutherfurd KJ. Viability and dose-response studies on the effects of the immunoenhancing lactic acid bacterium Lactobacillus rhamnosus in mice. *British Journal of Nutrition.* 2001; 86(2):285-9.

[69] Gill HS, Rutherfurd KJ, Cross ML, Gopal PK. Enhancement of immunity in the elderly by dietary supplementation with the probiotic Bifidobacterium lactis HN019. *American Journal of Clinical Nutrition.* 2001; 74(6):833-9.

[70] O'Mahony L, Feeney M, O'Halloran S, Murphy L, Kiely B, Fitzgibbon J, et al. Probiotic impact on microbial flora, inflammation and tumour development in IL-10 knockout mice. *Alimentary Pharmacology and Therapeutics.* 2001; 15(8):1219-25.

[71] Thompson C, McCarter YS, Krause PJ, Herson VC. Lactobacillus acidophilus sepsis in a neonate. *Journal of perinatology: official journal of the California Perinatal Association.* 2001; 21(4):258-60.

[72] Kalliomaki M, Salminen S, Arvilommi H, Kero P, Koskinen P, Isolauri E. Probiotics in primary prevention of atopic disease: a randomised placebo-controlled trial. *Lancet.* 2001; 357(9262):1076-9.

[73] Avlami A, Kordossis T, Vrizidis N, Sipsas NV. Lactobacillus rhamnosus endocarditis complicating colonoscopy. *The Journal of Infection.* 2001; 42(4):283-5.

[74] Barton LL, Rider ED, Coen RW. Bacteremic infection with Pediococcus: vancomycin-resistant opportunist. *Pediatrics.* 2001; 107(4):775-6.

[75] von der Weid T, Bulliard C, Schiffrin EJ. Induction by a lactic acid bacterium of a population of CD4(+) T cells with low proliferative capacity that produce transforming growth factor beta and interleukin-10. *Clinical and Diagnostic Laboratory Immunology*. 2001; 8(4):695-701.

[76] Gill HS, Rutherfurd KJ, Cross ML. Dietary probiotic supplementation enhances natural killer cell activity in the elderly: an investigation of age-related immunological changes. *Journal of Clinical Immunology*. 2001; 21(4):264-71.

[77] Liu Q, Nobaek S, Adawi D, Mao Y, Wang M, Molin G, et al. Administration of Lactobacillus plantarum 299v reduces side-effects of external radiation on colon anastomotic healing in an experimental model. *Colorectal Disease*. 2001; 3(4):245-52.

[78] Sheih YH, Chiang BL, Wang LH, Liao CK, Gill HS. Systemic immunity-enhancing effects in healthy subjects following dietary consumption of the lactic acid bacterium Lactobacillus rhamnosus HN001. *Journal of the American College of Nutrition*. 2001; 20(2 Suppl):149-56.

[79] Ulisse S, Gionchetti P, D'Alo S, Russo FP, Pesce I, Ricci G, et al. Expression of cytokines, inducible nitric oxide synthase, and matrix metalloproteinases in pouchitis: effects of probiotic treatment. *The American Journal of Gastroenterology*. 2001; 96(9):2691-9. Epub 2001/09/25.

[80] Madsen K, Cornish A, Soper P, McKaigney C, Jijon H, Yachimec C, et al. Probiotic bacteria enhance murine and human intestinal epithelial barrier function. *Gastroenterology*. 2001; 121(3):580-91.

[81] Reid G, Burton J. Use of Lactobacillus to prevent infection by pathogenic bacteria. *Microbes and Infection*. 2002; 4(3):319-24.

[82] Christensen HR, Frokiaer H, Pestka JJ. Lactobacilli differentially modulate expression of cytokines and maturation surface markers in murine dendritic cells. *Journal of Immunology*. 2002; 168(1): 171-8.

[83] Salminen MK, Tynkkynen S, Rautelin H, Saxelin M, Vaara M, Ruutu P, et al. Lactobacillus bacteremia during a rapid increase in probiotic use of Lactobacillus rhamnosus GG in Finland. *Clinical Infectious Diseases*. 2002; 35(10):1155-60.

[84] Pochard P, Gosset P, Grangette C, Andre C, Tonnel AB, Pestel J, et al. Lactic acid bacteria inhibit TH2 cytokine production by mononuclear cells from allergic patients. *Journal of Allergy and Clinical Immunology*. 2002; 110(4):617-23.

[85] McCracken VJ, Chun T, Baldeon ME, Ahrne S, Molin G, Mackie RI, et al. TNF-alpha sensitizes HT-29 colonic epithelial cells to intestinal lactobacilli. *Experimental Biology and Medicine*. 2002; 227(8):665-70.

[86] Cukrowska B, LodInova-ZadnIkova R, Enders C, Sonnenborn U, Schulze J, Tlaskalova-Hogenova H. Specific proliferative and antibody responses of premature infants to intestinal colonization with nonpathogenic probiotic E. coli strain Nissle 1917. *Scandinavian Journal of Immunology*. 2002; 55(2):204-9.

[87] He F, Morita H, Ouwehand AC, Hosoda M, Hiramatsu M, Kurisaki J, et al. Stimulation of the secretion of pro-inflammatory cytokines by Bifidobacterium strains. *Microbiology and Immunology*. 2002; 46(11):781-5.

[88] Perdigon G, Maldonado Galdeano C, Valdez JC, Medici M. Interaction of lactic acid bacteria with the gut immune system. *European Journal of Clinical Nutrition*. 2002; 56 Suppl 4:S21-6.

[89] Cross ML, Stevenson LM, Gill HS. Anti-allergy properties of fermented foods: an important immunoregulatory mechanism of lactic acid bacteria? *International Immunopharmacology*. 2001; 1(5):891-901.

[90] Borruel N, Carol M, Casellas F, Antolin M, de Lara F, Espin E, et al. Increased mucosal tumour necrosis factor alpha production in Crohn's disease can be downregulated ex vivo by probiotic bacteria. *Gut*. 2002; 51(5):659-64.

[91] Schultz M, Veltkamp C, Dieleman LA, Grenther WB, Wyrick PB, Tonkonogy SL, et al. Lactobacillus plantarum 299V in the treatment and prevention of spontaneous colitis in interleukin-10-deficient mice. *Inflammatory Bowel Diseases*. 2002; 8(2):71-80.

[92] Borruel N, Casellas F, Antolin M, Llopis M, Carol M, Espiin E, et al. Effects of nonpathogenic bacteria on cytokine secretion by human intestinal mucosa. *The American Journal of Gastroenterology*. 2003; 98(4):865-70.

[93] McCarthy J, O'Mahony L, O'Callaghan L, Sheil B, Vaughan EE, Fitzsimons N, et al. Double blind, placebo controlled trial of two probiotic strains in interleukin 10 knockout mice and mechanistic link with cytokine balance. *Gut*. 2003; 52(7):975-80.

[94] Riquelme AJ, Calvo MA, Guzman AM, Depix MS, Garcia P, Perez C, et al. Saccharomyces cerevisiae fungemia after Saccharomyces boulardii treatment in immunocompromised patients. *Journal of Clinical Gastroenterology*. 2003; 36(1):41-3.

[95] Lammers KM, Brigidi P, Vitali B, Gionchetti P, Rizzello F, Caramelli E, et al. Immunomodulatory effects of probiotic bacteria DNA: IL-1 and IL-10 response in human peripheral blood mononuclear cells. *FEMS Immunology and Medical Microbiology.* 2003; 38(2):165-72.

[96] Benyacoub J, Czarnecki-Maulden GL, Cavadini C, Sauthier T, Anderson RE, Schiffrin EJ, et al. Supplementation of food with Enterococcus faecium (SF68) stimulates immune functions in young dogs. *The Journal of Nutrition.* 2003; 133(4):1158-62.

[97] Braat H, van den Brande J, van Tol E, Hommes D, Peppelenbosch M, van Deventer S. Lactobacillus rhamnosus induces peripheral hyporesponsiveness in stimulated CD4+ T cells via modulation of dendritic cell function. *American Journal of Clinical Nutrition.* 2004; 80(6):1618-25.

[98] Jijon H, Backer J, Diaz H, Yeung H, Thiel D, McKaigney C, et al. DNA from probiotic bacteria modulates murine and human epithelial and immune function. *Gastroenterology.* 2004; 126(5):1358-73.

[99] Kunz AN, Noel JM, Fairchok MP. Two cases of Lactobacillus bacteremia during probiotic treatment of short gut syndrome. *Journal of Pediatric Gastroenterology and Nutrition.* 2004; 38(4):457-8.

[100] Panigrahi A, Kiron V, Kobayashi T, Puangkaew J, Satoh S, Sugita H. Immune responses in rainbow trout Oncorhynchus mykiss induced by a potential probiotic bacteria Lactobacillus rhamnosus JCM 1136. *Veterinary Immunology and Immunopathology.* 2004; 102(4):379-88.

[101] Schultz M, Strauch UG, Linde HJ, Watzl S, Obermeier F, Gottl C, et al. Preventive effects of Escherichia coli strain Nissle 1917 on acute and chronic intestinal inflammation in two different murine models of colitis. *Clinical and diagnostic laboratory immunology.* 2004; 11(2):372-8.

[102] Wehkamp J, Harder J, Wehkamp K, Wehkamp-von Meissner B, Schlee M, Enders C, et al. NF-kappaB- and AP-1-mediated induction of human beta defensin-2 in intestinal epithelial cells by Escherichia coli Nissle 1917: a novel effect of a probiotic bacterium. *Infection and Immunity.* 2004; 72(10):5750-8.

[103] Pohjavuori E, Viljanen M, Korpela R, Kuitunen M, Tiittanen M, Vaarala O, et al. Lactobacillus GG effect in increasing IFN-gamma production in infants with cow's milk allergy. *Journal of Allergy and Clinical Immunology.* 2004; 114(1):131-6.

[104] Chapat L, Chemin K, Dubois B, Bourdet-Sicard R, Kaiserlian D. Lactobacillus casei reduces CD8+ T cell-mediated skin inflammation. *European Journal of Immunology.* 2004; 34(9):2520-8.

[105] Hart AL, Lammers K, Brigidi P, Vitali B, Rizzello F, Gionchetti P, et al. Modulation of human dendritic cell phenotype and function by probiotic bacteria. *Gut.* 2004; 53(11):1602-9.
[106] Baillon ML, Marshall-Jones ZV, Butterwick RF. Effects of probiotic Lactobacillus acidophilus strain DSM13241 in healthy adult dogs. *American Journal of Veterinary Research.* 2004; 65(3):338-43.
[107] Fujiwara D, Inoue S, Wakabayashi H, Fujii T. The anti-allergic effects of lactic acid bacteria are strain dependent and mediated by effects on both Th1/Th2 cytokine expression and balance. *International Archives of Allergy and Immunology.* 2004; 135(3):205-15.
[108] Gerbitz A, Schultz M, Wilke A, Linde HJ, Scholmerich J, Andreesen R, et al. Probiotic effects on experimental graft-versus-host disease: let them eat yogurt. *Blood.* 2004; 103(11):4365-7.
[109] Land MH, Rouster-Stevens K, Woods CR, Cannon ML, Cnota J, Shetty AK. Lactobacillus sepsis associated with probiotic therapy. *Pediatrics.* 2005; 115(1):178-81.
[110] Sauter SN, Allenspach K, Gaschen F, Grone A, Ontsouka E, Blum JW. Cytokine expression in an ex vivo culture system of duodenal samples from dogs with chronic enteropathies: modulation by probiotic bacteria. *Domestic Animal Endocrinology.* 2005; 29(4):605-22.
[111] Lan JG, Cruickshank SM, Singh JC, Farrar M, Lodge JP, Felsburg PJ, et al. Different cytokine response of primary colonic epithelial cells to commensal bacteria. *World Journal of Gastroenterology.* 2005; 11(22):3375-84.
[112] Pena JA, Rogers AB, Ge Z, Ng V, Li SY, Fox JG, et al. Probiotic Lactobacillus spp. diminish Helicobacter hepaticus-induced inflammatory bowel disease in interleukin-10-deficient mice. *Infection and Immunity. 2005; 73(2):912-20.*
[113] *Ukena SN,* Westendorf AM, Hansen W, Rohde M, Geffers R, Coldewey S, et al. The host response to the probiotic Escherichia coli strain Nissle 1917: specific up-regulation of the proinflammatory chemokine MCP-1. *BMC Medical Genetics.* 2005; 6:43.
[114] Fernandez MF, Boris S, Barbes C. Safety evaluation of Lactobacillus delbrueckii subsp. lactis UO 004, a probiotic bacterium. *Research in Microbiology.* 2005; 156(2):154-60.
[115] Viljanen M, Pohjavuori E, Haahtela T, Korpela R, Kuitunen M, Sarnesto A, et al. Induction of inflammation as a possible mechanism of probiotic effect in atopic eczema-dermatitis syndrome. *Journal of Allergy and Clinical Immunology.* 2005; 115(6):1254-9.

[116] Westendorf AM, Gunzer F, Deppenmeier S, Tapadar D, Hunger JK, Schmidt MA, et al. Intestinal immunity of Escherichia coli NISSLE 1917: a safe carrier for therapeutic molecules. *FEMS Immunology and Medical Microbiology.* 2005; 43(3):373-84.

[117] Olivares M, Diaz-Ropero MP, Gomez N, Lara-Villoslada F, Sierra S, Maldonado JA, et al. The consumption of two new probiotic strains, Lactobacillus gasseri CECT 5714 and Lactobacillus coryniformis CECT 5711, boosts the immune system of healthy humans. *International Microbiology.* 2006; 9(1):47-52.

[118] Vinderola G, Perdigon G, Duarte J, Farnworth E, Matar C. Effects of the oral administration of the exopolysaccharide produced by Lactobacillus kefiranofaciens on the gut mucosal immunity. *Cytokine.* 2006; 36(5-6):254-60.

[119] Kim YG, Ohta T, Takahashi T, Kushiro A, Nomoto K, Yokokura T, et al. Probiotic Lactobacillus casei activates innate immunity via NF-kappaB and p38 MAP kinase signaling pathways. *Microbes and Infection.* 2006; 8(4):994-1005.

[120] Cukovic-Cavka S, Likic R, Francetic I, Rustemovic N, Opacic M, Vucelic B. Lactobacillus acidophilus as a cause of liver abscess in a NOD2/CARD15-positive. *Digestion.* 2006; 73(2-3):107-10.

[121] Duncker SC, Lorentz A, Schroeder B, Breves G, Bischoff SC. Effect of orally administered probiotic E. coli strain Nissle 1917 on intestinal mucosal immune cells of healthy young pigs. *Veterinary Immunology and Immunopathology.* 2006; 111(3-4):239-50.

[122] Marshall-Jones ZV, Baillon ML, Croft JM, Butterwick RF. Effects of Lactobacillus acidophilus DSM13241 as a probiotic in healthy adult cats. *American Journal of Veterinary Research.* 2006; 67(6):1005-12.

[123] Voltan S, Castagliuolo I, Elli M, Longo S, Brun P, D'Inca R, et al. Aggregating phenotype in Lactobacillus crispatus determines intestinal colonization and TLR2 and TLR4 modulation in murine colonic mucosa. *Clinical and Vaccine Immunology.* 2007; 14(9):1138-48.

[124] Galdeano CM, Perdigon G. The probiotic bacterium Lactobacillus casei induces activation of the gut mucosal immune system through innate immunity. *Clinical and Vaccine Immunology.* 2006;1 3(2):219-26.

[125] Castellazzi AM, Valsecchi C, Montagna L, Malfa P, Ciprandi G, Avanzini MA, et al. In vitro activation of mononuclear cells by two probiotics: Lactobacillus paracasei I 1688, Lactobacillus salivarius I 1794, and their mixture (PSMIX). *Immunological Investigations.* 2007; 36(4):413-21.

[126] Scharek L, Altherr BJ, Tolke C, Schmidt MF. Influence of the probiotic Bacillus cereus var. toyoi on the intestinal immunity of piglets. *Veterinary Immunology and Immunopathology*. 2007; 120(3-4):136-47.

[127] Nerstedt A, Nilsson EC, Ohlson K, Hakansson J, Thomas Svensson L, Lowenadler B, et al. Administration of Lactobacillus evokes coordinated changes in the intestinal expression profile of genes regulating energy homeostasis and immune phenotype in mice. *British Journal of Nutrition*. 2007; 97(6):1117-27.

[128] Ozkan TB, Sahin E, Erdemir G, Budak F. Effect of Saccharomyces boulardii in children with acute gastroenteritis and its relationship to the immune response. *Journal of International Medical Research*. 2007; 35(2):201-12.

[129] Niers LE, Hoekstra MO, Timmerman HM, van Uden NO, de Graaf PM, Smits HH, et al. Selection of probiotic bacteria for prevention of allergic diseases: immunomodulation of neonatal dendritic cells. *Clinical and Experimental Immunology*. 2007; 149(2):344-52.

[130] Foligne B, Nutten S, Grangette C, Dennin V, Goudercourt D, Poiret S, et al. Correlation between in vitro and in vivo immunomodulatory properties of lactic acid bacteria. *World Journal of Gastroenterology*. 2007; 13(2):236-43.

[131] Vinderola G, Matar C, Perdigon G. Milk fermentation products of L. helveticus R389 activate calcineurin as a signal to promote gut mucosal immunity. *BMC Immunology*. 2007; 8:19.

[132] Chuang L, Wu KG, Pai C, Hsieh PS, Tsai JJ, Yen JH, et al. Heat-killed cells of lactobacilli skew the immune response toward T helper 1 polarization in mouse splenocytes and dendritic cell-treated T cells. *Journal of Agricultural and Food Chemistry*. 2007; 55(26):11080-6.

[133] Lara-Villoslada F, Sierra S, Martin R, Delgado S, Rodriguez JM, Olivares M, et al. Safety assessment of two probiotic strains, Lactobacillus coryniformis CECT5711 and Lactobacillus gasseri CECT5714. *Journal of Applied Microbiology*. 2007; 103(1):175-84.

[134] Matsuzaki T, Takagi A, Ikemura H, Matsuguchi T, Yokokura T. Intestinal microflora: probiotics and autoimmunity. *The Journal of Nutrition*. 2007; 137(3 Suppl 2):798s-802s.

[135] Arslanoglu S, Moro GE, Schmitt J, Tandoi L, Rizzardi S, Boehm G. Early dietary intervention with a mixture of prebiotic oligosaccharides reduces the incidence of allergic manifestations and infections during the first two years of life. *The Journal of Nutrition*. 2008; 138(6):1091-5.

[136] Peluso I FD, Caruso R, Stolfi C, Caprioli F, Fantini MC, Caspani G, Grossi E, Di Iorio L, Paone FM, Pallone F, Monteleone G. Lactobacillus paracasei subsp. paracasei B21060 Suppresses Human T-Cell Proliferation. *Infection and Immunity*. 2007; 75(4):1730-7.

[137] Klein A, Friedrich U, Vogelsang H, Jahreis G. Lactobacillus acidophilus 74-2 and Bifidobacterium animalis subsp lactis DGCC 420 modulate unspecific cellular immune response in healthy adults. *European Journal of Clinical Nutrition*. 2008; 62(5):584-93.

[138] Feighery LM, Smith P, O'Mahony L, Fallon PG, Brayden DJ. Effects of Lactobacillus salivarius 433118 on intestinal inflammation, immunity status and in vitro colon function in two mouse models of inflammatory bowel disease. *Digestive Diseases and Sciences*. 2008; 53(9):2495-506.

[139] Imaoka A, Shima T, Kato K, Mizuno S, Uehara T, Matsumoto S, et al. Anti-inflammatory activity of probiotic Bifidobacterium: enhancement of IL-10 production in peripheral blood mononuclear cells from ulcerative colitis patients and inhibition of IL-8 secretion in HT-29 cells. *World Journal of Gastroenterology*. 2008; 14(16):2511-6.

[140] West CE, Gothefors L, Granstrom M, Kayhty H, Hammarstrom ML, Hernell O. Effects of feeding probiotics during weaning on infections and antibody responses to diphtheria, tetanus and Hib vaccines. *Pediatric Allergy and Immunology*. 2008; 19(1):53-60.

[141] Vidal K, Benyacoub J, Moser M, Sanchez-Garcia J, Serrant P, Segura-Roggero I, et al. Effect of Lactobacillus paracasei NCC2461 on antigen-specific T-cell mediated immune responses in aged mice. *Rejuvenation Research*. 2008; 11(5):957-64.

[142] Paturi G, Phillips M, Kailasapathy K. Effect of probiotic strains Lactobacillus acidophilus LAFTI L10 and Lactobacillus paracasei LAFTI L26 on systemic immune functions and bacterial translocation in mice. *Journal of Food Protection*. 2008; 71(4):796-801.

[143] Tsai YT, Cheng PC, Fan CK, Pan TM. Time-dependent persistence of enhanced immune response by a potential probiotic strain Lactobacillus paracasei subsp. paracasei NTU 101. *International Journal of Food Microbiology*. 2008; 128(2):219-25.

[144] Talebi A, Amirzadeh B, Mokhtari B, Gahri H. Effects of a multi-strain probiotic (PrimaLac) on performance and antibody responses to Newcastle disease virus and infectious bursal disease virus vaccination in broiler chickens. Avian pathology: *Journal of the WVPA*. 2008; 37(5):509-12.

[145] Reyes-Becerril M, Salinas I, Cuesta A, Meseguer J, Tovar-Ramirez D, Ascencio-Valle F, et al. Oral delivery of live yeast Debaryomyces hansenii modulates the main innate immune parameters and the expression of immune-relevant genes in the gilthead seabream (Sparus aurata L.). *Fish & Shellfish Immunology*. 2008; 25(6):731-9.
[146] Ghadimi D, Folster-Holst R, de Vrese M, Winkler P, Heller KJ, Schrezenmeir J. Effects of probiotic bacteria and their genomic DNA on TH1/TH2-cytokine production by peripheral blood mononuclear cells (PBMCs) of healthy and allergic subjects. *Immunobiology*. 2008; 213(8):677-92.
[147] Candela M, Perna F, Carnevali P, Vitali B, Ciati R, Gionchetti P, et al. Interaction of probiotic Lactobacillus and Bifidobacterium strains with human intestinal epithelial cells: Adhesion properties, competition against enteropathogens and modulation of IL-8 production. *International Journal of Food Microbiology*. 2008; 125(3):286-92.
[148] Takagi A, Ikemura H, Matsuzaki T, Sato M, Nomoto K, Morotomi M, et al. Relationship between the in vitro response of dendritic cells to Lactobacillus and prevention of tumorigenesis in the mouse. *Journal of Gastroenterology*. 2008; 43(9):661-9.
[149] Kekkonen RA, Lummela N, Karjalainen H, Latvala S, Tynkkynen S, Jarvenpaa S, et al. Probiotic intervention has strain-specific anti-inflammatory effects in healthy adults. *World Journal of Gastroenterology*. 2008; 14(13):2029-36.
[150] Paineau D, Carcano D, Leyer G, Darquy S, Alyanakian MA, Simoneau G, et al. Effects of seven potential probiotic strains on specific immune responses in healthy adults: a double-blind, randomized, controlled trial. *FEMS Immunology and Medical Microbiology*. 2008; 53(1):107-13.
[151] Ya T, Zhang Q, Chu F, Merritt J, Bilige M, Sun T, et al. Immunological evaluation of Lactobacillus casei Zhang: a newly isolated strain from koumiss in Inner Mongolia, China. *BMC Immunology*. 2008; 9:68.
[152] Maassen CB, Claassen E. Strain-dependent effects of probiotic lactobacilli on EAE autoimmunity. *Vaccine*. 2008; 26(17):2056-7.
[153] Nandakumar NS, Pugazhendhi S, Madhu Mohan K, Jayakanthan K, Ramakrishna BS. Effect of Vibrio cholerae on chemokine gene expression in HT29 cells and its modulation by Lactobacillus GG. *Scandinavian Journal of Immunology*. 2009; 69(3):181-7.
[154] Mileti E, Matteoli G, Iliev ID, Rescigno M. Comparison of the immunomodulatory properties of three probiotic strains of Lactobacilli

using complex culture systems: prediction for in vivo efficacy. *PLoS One.* 2009; 4(9):e7056.

[155] Hemmerling A, Harrison W, Schroeder A, Park J, Korn A, Shiboski S, et al. Phase 1 dose-ranging safety trial of Lactobacillus crispatus CTV-05 for the prevention of bacterial vaginosis. *Sexually Transmitted Diseases.* 2009; 36(9):564-9.

[156] Yang HY, Liu SL, Ibrahim SA, Zhao L, Jiang JL, Sun WF, et al. Oral administration of live Bifidobacterium substrains isolated from healthy centenarians enhanced immune function in BALB/c mice. *Nutrition Research.* 2009; 29(4):281-9.

[157] Martinez-Canavate A, Sierra S, Lara-Villoslada F, Romero J, Maldonado J, Boza J, et al. A probiotic dairy product containing L. gasseri CECT5714 and L. coryniformis CECT5711 induces immunological changes in children suffering from allergy. *Pediatric Allergy Immunology.* 2009; 20(6):592-600.

[158] Bickert T, Trujillo-Vargas CM, Duechs M, Wohlleben G, Polte T, Hansen G, et al. Probiotic Escherichia coli Nissle 1917 suppresses allergen-induced Th2 responses in the airways. *International Archives Allergy Immunology.* 2009; 149(3):219-30.

[159] Wang A, Yu H, Gao X, Li X, Qiao S. Influence of Lactobacillus fermentum I5007 on the intestinal and systemic immune responses of healthy and E. coli challenged piglets. *Antonie Van Leeuwenhoek.* 2009; 96(1):89-98.

[160] Ouwehand AC, Tiihonen K, Saarinen M, Putaala H, Rautonen N. Influence of a combination of Lactobacillus acidophilus NCFM and lactitol on healthy elderly: intestinal and immune parameters. *British Journal of Nutrition.* 2009; 101(3):367-75.

[161] Baron M. A patented strain of Bacillus coagulans increased immune response to viral challenge. *Postgraduate Medicine.* 2009; 121(2):114-8.

[162] Tseng DY, Ho PL, Huang SY, Cheng SC, Shiu YL, Chiu CS, et al. Enhancement of immunity and disease resistance in the white shrimp, Litopenaeus vannamei, by the probiotic, Bacillus subtilis E20. *Fish & Shellfish Immunology.* 2009; 26(2):339-44.

[163] Cammarota M, De Rosa M, Stellavato A, Lamberti M, Marzaioli I, Giuliano M. In vitro evaluation of Lactobacillus plantarum DSMZ 12028 as a probiotic: emphasis on innate immunity. *International Journal of Food Microbiology.* 2009; 135(2):90-8.

[164] Lombardo LV, A.; Blanco, I. Clinical evaluation of Lactobacillus paracasei subsp. paracasei F19 with gluco-oligosaccharides in the short-

term treatment of irritable bowel syrndrome. *Microbiol Ecology Health Diseases*. 2009; 21:28-32.

[165] Nishimura I, Igarashi T, Enomoto T, Dake Y, Okuno Y, Obata A. Clinical efficacy of halophilic lactic acid bacterium Tetragenococcus halophilus Th221 from soy sauce moromi for perennial allergic rhinitis. *Allergology International*. 2009; 58(2):179-85.

[166] Arribas B, Rodriguez-Cabezas ME, Camuesco D, Comalada M, Bailon E, Utrilla P, et al. A probiotic strain of Escherichia coli, Nissle 1917, given orally exerts local and systemic anti-inflammatory effects in lipopolysaccharide-induced sepsis in mice. *British Journal of Pharmacology*. 2009; 157(6):1024-33.

[167] Jain S, Yadav H, Sinha PR, Marotta F. Modulation of cytokine gene expression in spleen and Peyer's patches by feeding dahi containing probiotic Lactobacillus casei in mice. *Journal of Digestive Diseases*. 2009; 10(1):49-54.

[168] West CE, Hammarstrom ML, Hernell O. Probiotics during weaning reduces the incidence of eczema. *Pediatric Allergy and Immunology*. 2009; 20(5):430-7.

[169] Vlieger AM, Robroch A, van Buuren S, Kiers J, Rijkers G, Benninga MA, et al. Tolerance and safety of Lactobacillus paracasei ssp. paracasei in combination with Bifidobacterium animalis ssp. lactis in a prebiotic-containing infant formula: a randomised controlled trial. *British Journal of Nutrition*. 2009; 102(6):869-75.

[170] Simrén M OL, Olsson J, Svensson U, Ohlson K, Posserud I, Strid H. Clinical trial: the effects of a fermented milk containing three probiotic bacteria in patients with irritable bowel syndrome - a randomized, double-blind, controlled study. *Alimentary Pharmacology and Therapeutics*. 2010; 31(2):218-27.

[171] Sierra S, Lara-Villoslada F, Sempere L, Olivares M, Boza J, Xaus J. Intestinal and immunological effects of daily oral administration of Lactobacillus salivarius CECT5713 to healthy adults. *Anaerobe*. 2010; 16(3):195-200.

[172] Zhang L, Xu YQ, Liu HY, Lai T, Ma JL, Wang JF, et al. Evaluation of Lactobacillus rhamnosus GG using an Escherichia coli K88 model of piglet diarrhoea: Effects on diarrhoea incidence, faecal microflora and immune responses. *Veterinary Microbiology*. 2010; 141(1-2):142-8.

[173] Maldonado J, Lara-Villoslada F, Sierra S, Sempere L, Gomez M, Rodriguez JM, et al. Safety and tolerance of the human milk probiotic

strain Lactobacillus salivarius CECT5713 in 6-month-old children. *Nutrition*. 2010; 26(11-12):1082-7.

[174] Zeuthen LH, Fink LN, Metzdorff SB, Kristensen MB, Licht TR, Nellemann C, et al. Lactobacillus acidophilus induces a slow but more sustained chemokine and cytokine response in naive foetal enterocytes compared to commensal Escherichia coli. *BMC Immunology*. 2010; 11:2.

[175] Generoso SV, Viana M, Santos R, Martins FS, Machado JA, Arantes RM, et al. Saccharomyces cerevisiae strain UFMG 905 protects against bacterial translocation, preserves gut barrier integrity and stimulates the immune system in a murine intestinal obstruction model. *Archives of Microbiology*. 2010; 192(6):477-84.

[176] Foligne B, Dewulf J, Vandekerckove P, Pignede G, Pot B. Probiotic yeasts: anti-inflammatory potential of various non-pathogenic strains in experimental colitis in mice. *World Journal of Gastroenterology*. 2010; 16(17):2134-45.

[177] Hemmerling A, Harrison W, Schroeder A, Park J, Korn A, Shiboski S, et al. Phase 2a study assessing colonization efficiency, safety, and acceptability of Lactobacillus crispatus CTV-05 in women with bacterial vaginosis. *Sexually Transmitted Diseases*. 2010; 37(12):745-50.

[178] Perez-Cano FJ, Dong H, Yaqoob P. In vitro immunomodulatory activity of Lactobacillus fermentum CECT5716 and Lactobacillus salivarius CECT5713: two probiotic strains isolated from human breast milk. *Immunobiology*. 2010; 215(12):996-1004.

[179] O'Mahony D, Murphy S, Boileau T, Park J, O'Brien F, Groeger D, et al. Bifidobacterium animalis AHC7 protects against pathogen-induced NF-kappaB activation in vivo. *BMC Immunology*. 2010; 11:63.

[180] Chiu CH, Cheng CH, Gua WR, Guu YK, Cheng W. Dietary administration of the probiotic, Saccharomyces cerevisiae P13, enhanced the growth, innate immune responses, and disease resistance of the grouper, Epinephelus coioides. *Fish & Shellfish Immunology*. 2010; 29(6):1053-9.

[181] Allen SJ, Jordan S, Storey M, Thornton CA, Gravenor M, Garaiova I, et al. Dietary supplementation with lactobacilli and bifidobacteria is well tolerated and not associated with adverse events during late pregnancy and early infancy. *The Journal of Nutrition*. 2010; 140(3):483-8.

[182] Gronbach K, Eberle U, Muller M, Olschlager TA, Dobrindt U, Leithauser F, et al. Safety of probiotic Escherichia coli strain Nissle

1917 depends on intestinal microbiota and adaptive immunity of the host. *Infection and Immunity*. 2010; 78(7):3036-46.

[183] Harikrishnan R, Balasundaram C, Heo MS. Lactobacillus sakei BK19 enriched diet enhances the immunity status and disease resistance to streptococcosis infection in kelp grouper, Epinephelus bruneus. *Fish & Shellfish Immunology*. 2010; 29(6):1037-43.

[184] Kobayashi T, Kato I, Nanno M, Shida K, Shibuya K, Matsuoka Y, et al. Oral administration of probiotic bacteria, Lactobacillus casei and Bifidobacterium breve, does not exacerbate neurological symptoms in experimental autoimmune encephalomyelitis. *Immunopharmacology and Immunotoxicology*. 2010; 32(1):116-24.

[185] Romanin D, Serradell M, Gonzalez Maciel D, Lausada N, Garrote GL, Rumbo M. Down-regulation of intestinal epithelial innate response by probiotic yeasts isolated from kefir. *International Journal of Food Microbiology*. 2010; 140(2-3):102-8.

[186] Kotani Y, Shinkai S, Okamatsu H, Toba M, Ogawa K, Yoshida H, et al. Oral intake of Lactobacillus pentosus strain b240 accelerates salivary immunoglobulin A secretion in the elderly: A randomized, placebo-controlled, double-blind trial. *Immunity & Ageing*. 2010; 7:11.

[187] Ligaarden SC, Axelsson L, Naterstad K, Lydersen S, Farup PG. A candidate probiotic with unfavourable effects in subjects with irritable bowel syndrome: a randomised controlled trial. *BMC Gastroenterology*. 2010; 10:16.

[188] Won TJ, Kim B, Song DS, Lim YT, Oh ES, Lee do I, et al. Modulation of Th1/Th2 balance by Lactobacillus strains isolated from Kimchi via stimulation of macrophage cell line J774A.1 in vitro. *Journal of Food Science*. 2011; 76(2):H55-61.

[189] Mane J, Pedrosa E, Loren V, Gassull MA, Espadaler J, Cune J, et al. A mixture of Lactobacillus plantarum CECT 7315 and CECT 7316 enhances systemic immunity in elderly subjects. A dose-response, double-blind, placebo-controlled, randomized pilot trial. *Nutricion Hospitalaria*. 2011; 26(1):228-35.

[190] Zhang J, Deng J, Wang Z, Che C, Li YF, Yang Q. Modulatory effects of Lactobacillus salivarius on intestinal mucosal immunity of piglets. *Current Microbiology*. 2011; 62(5):1623-31.

[191] dos Santos LM, Santos MM, de Souza Silva HP, Arantes RM, Nicoli JR, Vieira LQ. Monoassociation with probiotic Lactobacillus delbrueckii UFV-H2b20 stimulates the immune system and protects germfree mice

against Listeria monocytogenes infection. *Medical Microbiology and Immunology*. 2011; 200(1):29-38.

[192] Manzoni P, Lista G, Gallo E, Marangione P, Priolo C, Fontana P, et al. Routine Lactobacillus rhamnosus GG administration in VLBW infants: a retrospective, 6-year cohort study. *Early Human Development*. 2011; 87 Suppl 1:S35-8.

[193] Kingma SD, Li N, Sun F, Valladares RB, Neu J, Lorca GL. Lactobacillus johnsonii N6.2 stimulates the innate immune response through Toll-like receptor 9 in Caco-2 cells and increases intestinal crypt Paneth cell number in biobreeding diabetes-prone rats. *The Journal of Nutrition*. 2011; 141(6):1023-8.

[194] Lee J, Yun HS, Cho KW, Oh S, Kim SH, Chun T, et al. Evaluation of probiotic characteristics of newly isolated Lactobacillus spp.: immune modulation and longevity. *International Journal of Food Microbiology*. 2011; 148(2):80-6.

[195] Ringel-Kulka T, Palsson OS, Maier D, Carroll I, Galanko JA, Leyer G, et al. Probiotic bacteria Lactobacillus acidophilus NCFM and Bifidobacterium lactis Bi-07 versus placebo for the symptoms of bloating in patients with functional bowel disorders: a double-blind study. *Journal of Clinical Gastroenterology*. 2011; 45(6):518-25.

[196] Simakachorn N, Bibiloni R, Yimyaem P, Tongpenyai Y, Varavithaya W, Grathwohl D, et al. Tolerance, safety, and effect on the faecal microbiota of an enteral formula supplemented with pre- and probiotics in critically ill children. *Journal of Pediatric Gastroenterology and Nutrition*. 2011; 53(2):174-81.

[197] Hutt P, Koll P, Stsepetova J, Alvarez B, Mandar R, Krogh-Andersen K, et al. Safety and persistence of orally administered human Lactobacillus sp. strains in healthy adults. *Beneficial Microbes*. 2011; 2(1):79-90.

[198] Klein G. Antibiotic resistance and molecular characterization of probiotic and clinical Lactobacillus strains in relation to safety aspects of probiotics. *Foodborne Pathogens and Disease*. 2011; 8(2):267-81.

[199] Schiffer C, Lalanne AI, Cassard L, Mancardi DA, Malbec O, Bruhns P, et al. A strain of Lactobacillus casei inhibits the effector phase of immune inflammation. *Journal of Immunology* 2011; 187(5):2646-55.

[200] Shimosato T, Tomida K, Otani H. Effect of Lactobacillus pentosus ONRIC b0240 on intestinal IgA production in mice fed differing levels of protein. *Journal of Agricultural and Food Chemistry*. 2011; 59(6):2646-51.

[201] Mangell P, Thorlacius H, Syk I, Ahrne S, Molin G, Olsson C, et al. Lactobacillus plantarum 299v does not reduce enteric bacteria or bacterial translocation in patients undergoing colon resection. *Digestive Diseases and Sciences.* 2012; 57(7):1915-24.

[202] Holscher HD, Czerkies LA, Cekola P, Litov R, Benbow M, Santema S, et al. Bifidobacterium lactis Bb12 enhances intestinal antibody response in formula-fed infants: a randomized, double-blind, controlled trial. *Journal of Parenteral and Enteral Nutrition.* 2012; 36(1 Suppl):106s-17s.

[203] Gouriet F, Million M, Henri M, Fournier PE, Raoult D. Lactobacillus rhamnosus bacteremia: an emerging clinical entity. *European Journal of Clinical Microbiology & Infectious Disease*.s 2012; 31(9):2469-80.

[204] Maccaferri S, Klinder A, Brigidi P, Cavina P, Costabile A. Potential probiotic Kluyveromyces marxianus B0399 modulates the immune response in Caco-2 cells and peripheral blood mononuclear cells and impacts the human gut microbiota in an in vitro colonic model system. *Applied and Environmental Microbiology.* 2012; 78(4):956-64.

[205] Finamore A, Roselli M, Britti MS, Merendino N, Mengheri E. Lactobacillus rhamnosus GG and Bifidobacterium animalis MB5 induce intestinal but not systemic antigen-specific hyporesponsiveness in ovalbumin-immunized rats. *The Journal of Nutrition.* 2012; 142(2):375-81.

[206] Roos TB, de Lara AP, Dummer LA, Fischer G, Leite FP. The immune modulation of Bacillus cereus var. Toyoi in mice immunized with experimental inactivated Bovine Herpesvirus Type 5 vaccine. *Vaccine.* 2012; 30(12):2173-7.

[207] Fanning S, Hall LJ, Cronin M, Zomer A, MacSharry J, Goulding D, et al. Bifidobacterial surface-exopolysaccharide facilitates commensal-host interaction through immune modulation and pathogen protection. Proceedings of the National Academy of Sciences of the United States of America. 2012; 109(6):2108-13.

[208] Thomas CM, Hong T, van Pijkeren JP, Hemarajata P, Trinh DV, Hu W, et al. Histamine derived from probiotic Lactobacillus reuteri suppresses TNF via modulation of PKA and ERK signaling. *PLoS One.* 2012; 7(2):e31951.

[209] Plantinga TS, van Bergenhenegouwen J, Jacobs C, Joosten LA, van't Land B, Garssen J, et al. Modulation of Toll-like receptor ligands and Candida albicans-induced cytokine responses by specific probiotics. *Cytokine.* 2012; 59(1):159-65.

[210] Zhong Y, Huang J, Tang W, Chen B, Cai W. Effects of probiotics, probiotic DNA and the CpG oligodeoxynucleotides on ovalbumin-sensitized Brown-Norway rats via TLR9/NF-kappaB pathway. *FEMS Immunology and Medical Microbiology*. 2012; 66(1):71-82.
[211] Schabussova I, Hufnagl K, Tang ML, Hoflehner E, Wagner A, Loupal G, et al. Perinatal maternal administration of Lactobacillus paracasei NCC 2461 prevents allergic inflammation in a mouse model of birch pollen allergy. *PLoS One*. 2012; 7(7):40271.
[212] Dong H, Rowland I, Yaqoob P. Comparative effects of six probiotic strains on immune function in vitro. *British Journal of Nutrition*. 2012; 108(3):459-70.
[213] Karlsson M, Jass J. Lactobacilli differently regulate expression and secretion of CXCL8 in urothelial cells. *Beneficial Microbes*. 2012; 3(3):195-203.
[214] Messaoudi S, Madi A, Prevost H, Feuilloley M, Manai M, Dousset X, et al. In vitro evaluation of the probiotic potential of Lactobacillus salivarius SMXD51. *Anaerobe*. 2012; 18(6):584-9.
[215] Magrone T, Jirillo E. The interplay between the gut immune system and microbiota in health and disease: nutraceutical intervention for restoring intestinal homeostasis. *Current Pharmaceutical Design*. 2013; 19(7):1329-42.
[216] Yoon HS, Ju JH, Lee JE, Park HJ, Lee JM, Shin HK, et al. The probiotic Lactobacillus rhamnosus BFE5264 and Lactobacillus plantarum NR74 promote cholesterol efflux and suppress inflammation in THP-1 cells. *Journal of the Science of Food and Agriculture*. 2013; 93(4):781-7.
[217] Taverniti V, Stuknyte M, Minuzzo M, Arioli S, De Noni I, Scabiosi C, et al. S-layer protein mediates the stimulatory effect of Lactobacillus helveticus MIMLh5 on innate immunity. *Applied and Environmental Microbiology*. 2013; 79(4):1221-31.
[218] Carey CM, Kostrzynska M. Lactic acid bacteria and bifidobacteria attenuate the proinflammatory response in intestinal epithelial cells induced by Salmonella enterica serovar Typhimurium. *Canadian Journal of Microbiology*. 2013; 59(1):9-17.
[219] Biswas G, Korenaga H, Nagamine R, Takayama H, Kawahara S, Takeda S, et al. Cytokine responses in the Japanese pufferfish (Takifugu rubripes) head kidney cells induced with heat-killed probiotics isolated from the Mongolian dairy products. *Fish & Shellfish Immunology*. 2013; 34(5):1170-7.

[220] Liu Y, Fatheree NY, Dingle BM, Tran DQ, Rhoads JM. Lactobacillus reuteri DSM 17938 changes the frequency of Foxp3+ regulatory T cells in the intestine and mesenteric lymph node in experimental necrotizing enterocolitis. *PLoS One*. 2013; 8(2):e56547. Epub 2013/02/26.

[221] Sun K, Xie C, Xu D, Yang X, Tang J, Ji X. Lactobacillus isolates from healthy volunteers exert immunomodulatory effects on activated peripheral blood mononuclear cells. *Journal of Biomedical Research.* 2013; 27(2):116-26.

[222] Taverniti VG, S. The immunomodulatory properties of probiotic microorganisms beyond their viability (ghost probiotics: proposal of paraprobiotic concept. *Genes and Nutrition*. 2011; 6:261–74.

[223] Sanders MEG, F.; Guerrant, R.; Holt, P. R.; Quigley, E. M. M.; Sartor, B. R.; Sherman, P. M.; Mayer EA. An update on the use and investigation of probiotics in health and disease. *Gut.* 2013; 62:787-96.

[224] Marco MLT, S. Environmental factors influencing the efficacy of probiotic bacteria. *Current Opinion in Biotechnology*. 2013; 24:207-13.

[225] Saad ND, C.; Urdaci, M.; Schmitter; J. M.; Bressollier, P. An overview of the last advances in probiotic and prebiotic field. LWT - *Food Science and Technology*. 2013; 50:1-16.

[226] Besselink MG, van Santvoort HC, Buskens E, Boermeester MA, van Goor H, Timmerman HM, et al. Probiotic prophylaxis in predicted severe acute pancreatitis: a randomised. *Lancet.* 2008; 371(9613):651-9.

[227] Cross MLG, A.; Teilab, D.; Fray, L. M. Patterns of cytokine induction by Gram-positive and Gram-negative probiotic bacteria. *FEMS Immunology and Medical Microbiology*. 2004; 42:173-80.

Author Contact Information

Alessandro Di Cerbo, Ph.D.
Department of Surgery and Surgical Specialties
University of Modena and Reggio Emilia Medical School
Surgical Clinic, Modena, Italy
tel. +390594222483
e-mail. Alessandro811@hotmail.it

Beniamino Palmieri, M.D., Ph.D.
Department of Surgery and Surgical Specialties
University of Modena and Reggio Emilia Medical School
Surgical Clinic, Modena, Italy
tel. +390594222687
e-mail. palmieri@unimo.it

Maria Aponte, M.Sc.
Department of Agriculture
University of Naples "Federico II"
Portici (NA), Italy
tel. +390812539398
e-mail. aponte@unina.it

Index

D

G

H

I

N

O

Q

R

S

T

U

V

W

Y